OCEANOGRAPHY AND OCEAN ENGINEERING SERIES

MARINE PROTECTED AREAS

OCEANOGRAPHY AND OCEAN ENGINEERING SERIES

Ocean Engineering Research Advances
Alan I. Prescott (Editor)
2008. ISBN 1-60021-777-x

Coral Reefs: Biology, Threats and Restoration
Thomas B. Davin and Anna P. Brannet (Editors)
2009. ISBN 978-1-60692-104-3

Marine Phytoplankton
William T. Kersey and Samuel P. Munger (Editors)
2009. ISBN 978-1-60741-087-4

Marine Protected Areas
Floyd B. Mayr (Editor)
2010. ISBN 978-1-60741-549-7

OCEANOGRAPHY AND OCEAN ENGINEERING SERIES

MARINE PROTECTED AREAS

FLOYD B. MAYR
EDITOR

Nova Science Publishers, Inc.
New York

For permission to use material from this book please contact us:
Telephone 631-231-7269; Fax 631-231-8175
Web Site: http://www.novapublishers.com

LIBRARY OF CONGRESS CATALOGING-IN-PUBLICATION DATA

Marine protected areas / [edited by] Floyd B. Mayr.
p. cm.
Includes index.
ISBN 978-1-60741-549-7 (hardcover)
1. Marine protected areas--United States. I. Mayr, Floyd B. II. Upton, Harold F. (Harold Frank) III. Buck, Eugene H. IV. Vann, Adam. V. Upton, Harold F. (Harold Frank). Marine protected areas. VI. Vann, Adam. Marine portected areas (MPAs). VII. National Marine Protected Areas Center (U.S.) VIII. Framework for the National System of Marine Protected Areas of the United States of America.
QH91.75.U6M37 2009
333.95'6160973--dc22 2009032062

Published by Nova Science Publishers, Inc. ✢ New York

CONTENTS

PREFACE

This book explores the increasing impacts on the world's oceans from coastal and offshore development, overfishing, a changing climate, natural events, and other sources that are straining the health of marine ecosystems and the Great lakes. Impacts to these intricately balanced environments include declining fish populations, degradation of coral reefs and other vital habitats, threats to rare or endangered species, and loss of artifacts and resources that represent the diverse cultural heritage of the United States. In the United States and around the world, marine protected areas (MPAs) are increasingly recognized as an important and promising management tool for mitigating or buffering some of these impacts. This book examines the use of MPAs as a way to restore and maintain healthy marine and Great Lakes environments by contributing to the overall protection of critical marine habitats and resources.

This is an edited, excerpted and augmented edition of various government publications.

Chapter 1 - There continues to be congressional interest in limiting human activity in certain areas of the marine environment, as one response to mounting evidence of declining environmental quality and populations of living resources. The purposes of proposed additional limits would be both to stem declines and to permit the rehabilitation of these environments and populations. One method of implementing this concept is for Congress to designate areas where activities would be limited, often referred to as marine protected areas (MPAs). Translating the MPA approach into a national program, however, would require that Congress resolve many economic, ecological, and social dilemmas.

The complexity of creating a program is compounded by controversy over the uses that would be allowed, curtailed, or prohibited in MPAs; the purposes of a system of MPAs; and the location, size, and distribution of MPA units. One possible way to get past some of these complexities is to think of MPA designations as a form of zoning in the ocean. Experiences related to designating MPAs in other countries also may be instructive. However, questions have arisen about the effectiveness of administration and enforcement, the benefits and costs of MPAs, and the evaluation of outcomes at some sites.

Numerous marine sites have been designated by federal and state governments for some kind of protection. Perhaps the best-known federal sites are units in the National Marine Sanctuary System. The National Marine Sanctuaries Act authorizes the Secretary of Commerce to designate areas of marine and Great Lakes environments to protect cultural and natural resources. The Bush Administration also has supported the MPA concept. It designated the Papandnaumolcuäkea Marine National Monument (Northwestern Hawaiian

Islands Marine National Monument) in 2006 as one of the world's largest MPAs and has supported the activities of the National Marine Protected Areas Center in the National Oceanic and Atmospheric Administration. It has continued most of the Clinton Administration initiatives to coordinate protection of marine resources at designated sites, including implementing Executive Order 13158 (May 2000), which endorsed a comprehensive system of MPAs.

Additional actions by Congress would be needed to create an MPA system that could be characterized as integrated or comprehensive. Some issues that would likely be raised in congressional discussions include whether new legislation is desired or needed; what the basic characteristics of units in any MPA system should be; how MPAs might be used to resolve use conflicts; and whether adequate funding would be authorized and appropriated to both enforce the protected status and evaluate the ecological and social impacts of MPAs. Over the last three decades Congress has examined the concepts behind MPAs and experiences with protected areas as it considered appropriations and proposals to reauthorize coastal and marine resource protection laws. The 110th Congress has considered some MPA-related activities such as enhancing coral reef conservation, modifying boundaries and protections for three national marine sanctuaries, and reauthorizing the National Marine Sanctuary Act.

Chapter 2 - Recent events, including the release of the President's U.S. Ocean Action Plan and reports issued by the United States Commission on Ocean Policy and the Pew Oceans Commission, have prompted a reexamination of U.S. ocean policy and debate over an " ecosystem approach" to ocean resource management. One proposed mechanism for conserving ocean resources is the Marine Protected Area (MPA), conceptualized as a zoning system for the portions of the ocean under U.S. jurisdiction. This has been highlighted by the issuance of the Revised Draft Framework for Developing the National System of Marine Protected Areas (Draft Framework), issued on March 17, 2008, by the National Marine Protected Areas Center, part of the National Oceanic and Atmospheric Administration (NOAA). The introduction of H.R. 21 in the 110th Congress has also focused attention on the health of ocean resources.

The relative merits and the potentially negative consequences of such an MPA system have been widely discussed. Advocates of additional protection argue that a more comprehensive system as outlined in the Draft Framework should be established. Others argue that the current system is effectively managing ocean resources and that additional restrictions would be economically harmful.

Apart from the relative merits of each position, there is some question as to the applicability of current federal law to the oceans and whether new protections could be imposed administratively, without additional legislation. To some extent, regulatory authority depends upon the nature of the jurisdiction that the United States has claimed over various ocean resources vis-a-vis other nations and vis-a-vis the states. Consistent with international law, the United States claims jurisdiction over marine areas extending 200 nautical miles from its coast and has regulated resources in the zones composing this area under multiple legal authorities.

Several current laws which might provide authority for the creation of MPAs are aimed specifically at the ocean environment. The National Marine Sanctuary Program, established by the Marine Protection, Research and Sanctuaries Act, the Magnuson-Stevens Fishery

Conservation and Management Act, and the Coastal Zone Management Act each specifically contemplate various levels and forms of aquatic resource protection.

Additionally, certain generally applicable laws, while primarily intended for use on land, would arguably support the designation of an MPA in some circumstances. Indeed, U.S. MPAs within the territorial seas have been established as national monuments, national parks, national wildlife areas, and, most recently, as a reserve via executive order.

This chapter outlines U.S. jurisdiction over ocean resources and analyzes the existing laws to assess their application to marine environments.

Chapter 3 - Increasing impacts on the world's oceans from coastal and offshore development, overfishing, a changing climate, natural events, and other sources are straining the health of marine ecosystems and the Great Lakes. Impacts to these intricately balanced environments include declining fish populations, degradation of coral reefs and other vital habitats, threats to rare or endangered species, and loss of artifacts and resources that represent the diverse cultural heritage of the United States. The effects of these losses are significant and jeopardize the social and economic fabric of the nation.

In the United States and around the world, marine protected areas (MPAs) are increasingly recognized as an important and promising management tool for mitigating or buffering some of these impacts. When used effectively and as a part of a broader ecosystem-based approach to management, MPAs can help to restore and maintain healthy marine and Great Lakes environments by contributing to the overall protection of critical marine habitats and resources. In this way, effective MPAs also can offer social and economic opportunities for current and future generations, such as tourism, biotechnology, fishing, education, and scientific research.

MPAs are designated and managed at all levels of government by a variety of agencies including parks, fisheries, wildlife, natural resource and historic resource departments, among others. U.S. MPAs have been established by well over 100 legal authorities, with some federal and state agencies managing more than one MPA program, each with its own legal purpose. There are approximately 1,700 existing MPAs in the United States that have been established by federal, state, territorial, and local governments to protect and conserve the nation's rich natural and cultural marine heritage and sustainable production resources. These MPAs have been designated to achieve a myriad of conservation objectives, ranging from conservation of biodiversity hotspots, to preservation of sunken historic vessels, to protection of spawning aggregations important to commercial and recreational fisheries. Similarly, the level of protection provided by these MPAs ranges from fully protected or no-take marine reserves to sites allowing multiple uses, including fishing, recreational, and industrial uses.

Recognizing the significant role that U.S. MPAs play in conserving marine heritage and sustainable use, and the lack of a national institution for comprehensive MPA planning, coordination, and support, Presidential Executive Order 13158 of May 26, 2000 (Order), found in Appendix D of this document, calls for the development of a National System of Marine Protected Areas (national system). The Order clearly calls for a national and not a federal system, and requires collaboration not only with other federal agencies, but also with coastal states and territories, tribes, Regional Fishery Management Councils, and other entities, as appropriate, including the MPA Federal Advisory Committee. The Order further specifies that the national system be scientifically based, comprehensive, and represent the nation's diverse marine ecosystems and natural and cultural resources.

To provide a blueprint for building the National System of MPAs, the Order calls for the development of a framework for a National System of MPAs and directs the establishment of a National MPA Center (MPA Center) within the National Oceanic and Atmospheric Administration (NOAA) to lead the system's development and implementation. This final *Framework for the National System of MPAs of the United States of America* (Framework) is the result of a multi-year development effort. The first draft Framework received over 11,000 comment submissions (composed of comments from 100 individual commenters and a petition from nearly 11,000 people) during its September 2006 to February 2007 public comment period. A second draft addressing these comments was published for public comment from March-May 2008, and received 34 public comment submissions. The MPA Federal Advisory Committee also provided two sets of recommendations on the Framework that have contributed significantly to its final form.

The Framework recognizes that U.S. MPA programs can achieve more efficient, effective conservation of the nation's important natural and cultural resources by working together rather than separately, and that many solutions require collaboration across programs with their own individual mandates, levels of government, and even international boundaries. It proposes a national system that is, initially, an assemblage of existing MPA sites, systems, and networks established and managed by federal, state, territorial, commonwealth, tribal, or local governments, acknowledging and building upon the contributions of these foundation programs. In addition, the Framework outlines collaborative, transparent processes for MPA programs at all levels of government to work together at regional, national, and international levels and with public participation to achieve common conservation objectives through comprehensive MPA planning; identification of enhanced or new MPAs that may be needed; and support for improved MPA science, stewardship, and effectiveness.

In: Marine Protected Areas
Editor: Floyd B. Mayr

ISBN: 978-1-60741-549-7

Chapter 1

MARINE PROTECTED AREAS: AN OVERVIEW

Harold F. Upton and Eugene H. Buck

SUMMARY

There continues to be congressional interest in limiting human activity in certain areas of the marine environment, as one response to mounting evidence of declining environmental quality and populations of living resources. The purposes of proposed additional limits would be both to stem declines and to permit the rehabilitation of these environments and populations. One method of implementing this concept is for Congress to designate areas where activities would be limited, often referred to as marine protected areas (MPAs). Translating the MPA approach into a national program, however, would require that Congress resolve many economic, ecological, and social dilemmas.

The complexity of creating a program is compounded by controversy over the uses that would be allowed, curtailed, or prohibited in MPAs; the purposes of a system of MPAs; and the location, size, and distribution of MPA units. One possible way to get past some of these complexities is to think of MPA designations as a form of zoning in the ocean. Experiences related to designating MPAs in other countries also may be instructive. However, questions have arisen about the effectiveness of administration and enforcement, the benefits and costs of MPAs, and the evaluation of outcomes at some sites.

Numerous marine sites have been designated by federal and state governments for some kind of protection. Perhaps the best-known federal sites are units in the National Marine Sanctuary System. The National Marine Sanctuaries Act authorizes the Secretary of Commerce to designate areas of marine and Great Lakes environments to protect cultural and natural resources. The Bush Administration also has supported the MPA concept. It designated the Papandnaumolcuäkea Marine National Monument (Northwestern Hawaiian Islands Marine National Monument) in 2006 as one of the world's largest MPAs and has supported the activities of the National Marine Protected Areas Center in the National Oceanic and Atmospheric Administration. It has continued most of the Clinton Administration initiatives to coordinate protection of marine resources at designated sites, including implementing Executive Order 13158 (May 2000), which endorsed a comprehensive system of MPAs.

Additional actions by Congress would be needed to create an MPA system that could be characterized as integrated or comprehensive. Some issues that would likely be raised in congressional discussions include whether new legislation is desired or needed; what the basic characteristics of units in any MPA system should be; how MPAs might be used to resolve use conflicts; and whether adequate funding would be authorized and appropriated to both enforce the protected status and evaluate the ecological and social impacts of MPAs. Over the last three decades Congress has examined the concepts behind MPAs and experiences with protected areas as it considered appropriations and proposals to reauthorize coastal and marine resource protection laws. The 110th Congress has considered some MPA-related activities such as enhancing coral reef conservation, modifying boundaries and protections for three national marine sanctuaries, and reauthorizing the National Marine Sanctuary Act.

Acknowledgment

This report was originally drafted by retired CRS specialist Jeffrey A. Zinn.

Many coastal and offshore ecosystems continue to be degraded by anthropogenic causes, despite efforts to control or limit them. The causes of degradation are numerous, and can include:

- pollutants;
- runoff (carrying sediment and chemicals) from land;
- coastal development;
- introduction of non-native or invasive species;
- overfishing and bycatch;
- habitat alteration; and
- rising sea level and climate change.

The public is becoming aware of degraded marine areas because of widely publicized incidents and trends, including a large seasonal "dead zone" in the Gulf of Mexico, the environmental effects of oil spills, population declines of many popular fish species to levels that can no longer sustain commercial or even recreational harvests, and deteriorating coral reefs (reef bleaching). Current approaches to managing resources in the marine environment often appear to be ineffective because of continuing population and environmental quality declines, thus prompting a search for alternatives.

Marine protected areas (MPAs) are generally defined as areas reserved by law or other effective means to protect part or all of the enclosed environment. Some observers, often including scientists and environmental advocates, recommend designating MPAs to achieve management and conservation goals. From their perspective, the designation of MPAs is not a panacea that responds to all causes of degradation, or leads to a quick recovery for all degraded environments, but in many cases they contend that MPAs are necessary for protecting and restoring the marine environment. Policy makers are looking at how this tool has worked, alternative ways that MPAs can be designed, and whether and how MPAs might be broadly applied. Little opposition has been expressed about the overall concept of establishing MPAs, but some of the more specific discussions about which uses would be limited or prohibited have been controversial. Oil and gas development, the fishing industry, and other marine industries have consistently expressed concerns with the use of MPAs. These industries question whether the use of MPAs would afford the proper balance between conservation and economic activities.

The 110th Congress has considered reauthorization of the National Marine Sanctuary Act (NMSA; 16 U.S.C. §§ 1431, et seq.) with introduction of H.R. 6537 and by holding several hearings, but no legislation has been passed. Many hold that NMSA comes closest to authorizing MPAs. It authorizes the National Oceanic and Atmospheric Administration (NOAA) to designate specific sites for comprehensive and coordinated management and conservation. However, some, especially environmentalists and many marine scientists, assert that a more comprehensive approach with stricter protection and more extensive protected areas is still needed.

This report identifies a number of issues related to establishing MPAs in the United States. It begins by defining the concept and administrative actions taken to provide spatial protection in marine areas. It then considers some of the key issues and potential benefits and costs of designating additional MPAs. It concludes by summarizing current congressional interest in the topic and legislation proposed during the 110th Congress. Existing federal laws related to the use of MPAs are summarized in the **Appendix.**

THE CONCEPT OF MARINE PROTECTED AREAS

The term MPA has been used to characterize spatial protection of marine resources, but ambiguities exist regarding the level of protection that qualifies an area as an MPA. Protection might be considered on a continuum — from no protection to complete protection where all human activities within an area are prohibited. Those areas afforded complete protection would certainly quality as MPAs, but areas closed to a specific use or to protect a species might not. Virtually all of U.S. waters are protected in some manner, but all U.S. waters are not considered MPAs. Often the definition depends on the context, such as the characteristics of the resources at risk and the extent of associated threats. The following section provides basic definitions of MPAs, but application of the concept is likely to evolve with improvements in scientific understanding of the marine environment and development of associated institutions.

Definition

MPAs have been defined in many ways. Definitions usually include three criteria: (1) geographically defined and bounded places; (2) approaches that manage systems rather than individual resources or species; and (3) programs that take a long-term perspective on resource management. The definition currently being used in this country, found in Section 2 of E.O. 13158 on Marine Protected Areas, defines MPAs as "any area of the marine environment that has been reserved by Federal, State, territorial, tribal, or local laws or regulations to provide lasting protection for part or all of the natural and cultural resources therein."

The E.O. 13158 definition is very similar to what is probably the most widely used defmition in the world, developed by the International Union for the Conservation of Nature (IUCN). It states that an MPA is "an area of intertidal or subtidal terrain, together with its overlying water and associated flora, fauna, historical and cultural features, which has been reserved by law or other effective means to protect part or all of the enclosed environment."[1] Each designated area must be a minimum of 1,000 hectares (2,471 acres). This definition is then applied to distinct categories of areas that reflect a wide range of possible management objectives. The IUCN specifies that sites meet this definition only when at least three-quarters of a designated MPA is managed for the primary category and management of the remaining area is not in conflict. The IUCN categories are:

Category 1: *Strict Nature Reserve/Wilderness Area*, managed mainly for science or wilderness protection.

- **Category la:** Strict nature reserve managed mainly for science.
- **Category lb:** Strict nature reserve managed mainly for wilderness protection.

Category 2: *National Park*, protected and managed mainly for ecosystem protection and recreation.

Category 3: *National Monument*, managed mainly for conservation of specific natural features.

Category 4: *Habitat Species Management Area*, managed mainly for conservation through management intervention.

Category 5: *Protected Landscape/Seascape*, managed mainly for landscape/seascape conservation and recreation.

Category 6: *Managed Resource Protected Area*, managed mainly for the sustainable use of natural ecosystems.

As one becomes more specific about possible goals and objectives for an MPA, the question of how MPAs would be defined quickly grows complex. They could involve state and federal jurisdiction, each with different goals and responsibilities. They could emphasize either some species (perhaps the commercially most valuable or the rarest ones) over others, or the general health of the entire ecosystem over individual components. Among the questions currently receiving the most attention are: (1) would certain activities or uses be automatically prohibited at all places designated as MPAs, such as automatically closing them to all resource extractive activities, or would a list of permitted and prohibited activities be developed for each designated area or category of areas; and (2) would environmental conditions or resource uses beyond the boundaries of MPAs be managed to limit adverse effects on resources within MPAs?

Definitions are one way to convey expectations about an MPA program. For example, one could conclude that most of the Gulf of Maine is protected by an impressive web of areas designated for a variety of reasons. However, a closer examination shows that most of these designations protect a single commercial fish species or group of species or limit a specified activity. Some want MPAs to be based on a broad definition to include designations where protections are limited to a few resources or uses, while others will want it to apply only to sites where resources are strictly protected. Between these two possibilities are many intermediate approaches. This debate over definition continues because the MPA concept has been used in many different ways. Several narrower terms are being used to describe types of places that provide high levels of protection, and they also have been used with different meanings. These places might also be called MPAs, although they are often expressed as if they were alternatives to MPAs. These terms include:

- *marine reserv,* where uses that remove resources are generally prohibited (these areas may also be called ecological reserves);
- *ocean wilderness*, like the terrestrial concept for wilderness areas on federal lands where no alterations or activities that leave lasting impacts are permitted, but low-impact recreational activities may be permitted;
- *fully protected marine area*, generally a "no-take" area where a wide variety of extractive and consumptive uses/activities are prohibited;
- *national marine sanctuary*, a specific designation created in federal legislation more than 30 years ago to ensure conservation and management for areas of special national significance;
- *marine managed area*, managing for multiple objectives, where protection is not the only, and may not even be the main, objective; and
- *marine park*, similar to the terrestrial concept for a park where recreational activities are allowed and resource conservation is also a goal of the designation.

ADMINISTRATIVE ACTIONS

According to the MPA Center definition, approximately 1,700 MPAs are managed by federal, state, and territorial agencies in U.S. waters.[2] These areas cover 34% of U.S. marine waters and vary widely in level of protection, restrictions on human uses, and overall purpose.[3] Less than 30% of MPAs are managed by federal agencies, but because of the large size of several federal MPAs, nearly 85% of the total MPA area in U.S. waters is under federal jurisdiction. About 90% of MPAs permit access and multiple use, including fishing. Most are permanent, providing protection throughout the year, and most have been established since 1970. NOAA manages MPAs for fisheries management (National Marine Fisheries Service), units of the National Estuarine Research Reserve System (Coastal Programs in National Ocean Service), and the National Marine Sanctuary Program (National Ocean Service). The Department of the Interior manages marine areas that are part of the National Wildlife Refuge System (Fish and Wildlife Service), National Parks (National Park Service), and national monuments (Fish and Wildlife Service and National Park Service). A number of federal laws and programs already exist, and they have been summarized in the **Appendix**.

It is thus apparent that the protection of marine resource areas through MPAs is not a new concept, but what is relatively new is the increasing interest in developing a coordinated, nationwide system of marine protected areas. This interest has been heightened by actions initiated by the Clinton Administration and continued by the Bush Administration. This concept has diffuse roots and has been most evident among scientists and some nongovernmental organizations.

President Clinton responded to growing concerns about marine resource degradation in several ways. He issued Executive Order 13089 on coral reef protection in June 1998 (1998 was also the internationally recognized "Year of the Ocean"). In 1999 and 2000, the Administration issued action plans calling for the federal government to work with state, territorial, and nongovernmental partners to expand and strengthen MPAs throughout the United States.[4] On May 26, 2000, President Clinton issued the Marine Protected Areas Executive Order, E.O. 13158. This order called for "strengthening and expanding the Nation's system of marine protected areas ... throughout the marine environment ... [to] enhance the conservation of our Nation's natural and cultural marine heritage and the ecologically and economically sustainable use of the marine environment for future generations."[5] More specifically, this order:

- aims to increase coordination and effectiveness of MPAs (but did not change existing MPAs or establish new MPAs);
- directs federal agencies to comply with existing regulations regarding MPAs (but did not alter existing regulations or authorities); and
- charges NOAA and the Department of the Interior with leading federal efforts by (1) creating a list of existing MPAs in the United States; (2) creating a national MPA webpage; (3) establishing a national MPA Center to provide tools and strategies for promoting MPA effectiveness; (4) establishing an MPA Advisory Committee to provide recommendations from stakeholders outside the federal government;[6] and (5) consulting with governmental and nongovernmental stakeholders.

On December 4, 2000, President Clinton issued E.O. 13178, creating the Northwestern Hawaiian Islands (NWHI) Coral Reef Ecosystem Reserve. This reserve, encompassing about 120,000 square miles (about 77 million acres), is the largest protected area ever established in the United States. Within the overall reserve, 15 reserve preservation areas encompassing about 6,200 square miles (nearly 4 million acres, or about 5% of the reserve) were designated where all consumptive or extractive activities are limited. NOAA managed the reserve under the authority of the NMSA.

President Bush has generally continued the initiatives started by the Clinton Administration, and has expanded on them in some cases. On June 4, 2001, after a review, Secretary of Commerce Donald L. Evans announced the retention of the Marine Protected Areas E.O. 13158. On June 15, 2006, President Bush established the Papandnaumokudkea Marine National Monument by proclamation under the Antiquities Act that encompasses the areas protected by the original NWHI reserve.

In August 2008, President Bush proposed protection for nearly 900,000 square miles in the Central and Western Pacific under the 1906 Antiquities Act (16 U.S.C. §§ 431-443). The Central Pacific areas would include waters adjacent to Wake, Howland, Baker and Jarvis islands; Johnson, Palmyra, and Rose Atolls; and Kingman reef. In the Western Pacific the area would include the waters around the northern islands of the Northern Mariana Islands, including the Mariana Trench. The Governor of Saipan (Northern Mariana Islands) and some Guam fishermen have opposed the Marianas Trench proposal while several environmental groups have voiced public support. The proposal is being reviewed by NOAA, the Interior Department, Defense Department, and the Council on Environmental Quality.

On September 26, 2008, President Bush amended E.O. 12962 to ensure that recreational fishing is "managed as a sustainable activity in national wildlife refuges, national parks, national monuments, national marine sanctuaries, marine protected areas, or any other relevant conservation or management areas under any Federal authority, consistent with applicable law." According to recreational interests, this action allays their concerns of being excluded from marine protected areas.[7]

MPA Center Activities

In 2000, the National MPA Center was created to implement E.O. 13158 "to develop a framework for a national system of MPAs, and to provide Federal, State, territorial, tribal, and local governments with the information, technologies, and strategies to support the system."[8] In 2003, the center established a 30-member MPA Advisory Committee that has provided broad representation of marine regions, including the Great Lakes and the U.S. territories. The committee's main purpose is to provide advice to the Secretaries of Commerce and the Interior on developing a national system of MPAs. In March of 2008, the center released a revised draft framework containing a set of recommendations for developing a national system of MPAs. The framework, developed after numerous workshops and advisory committee meetings, proposes guiding principles, goals and objectives, and definitions for a national system.[9]

The MPA Center has built a domestic inventory of federal, state, local, and tribal MPAs to inform the development of the national system called for in E.O. 13158.[10] The inventory may be used to provide information for environmental assessments; to lay a foundation for objective analysis in designing a national MPA system; and to provide a centralized source of information that can be used to help protect marine resources. The inventory contains information such as management authority, for nearly 1,700 sites.

Other initiatives of the center include a virtual library and education projects to increase knowledge and awareness about MPAs. The center provides educators with materials about topics that are relevant to MPAs, such as marine habitats and fisheries, through the MPA website and workshops. It is also leading a multi-year pilot program in California, Oregon, and Washington to try to improve approaches for designing and managing a system of MPAs at a regional level.

Other MPA Management and Coordination Efforts

While the center has coordinated many aspects of the federal interest in MPAs, other federal bodies and state governments have been active in managing and coordinating activities related to MPAs. The National Park Service Ocean Park Stewardship Action Plan highlights the establishment, in partnership with NOAA, other relevant agencies, and public and private entities, of a seamless system of ocean parks, sanctuaries, refuges, and reserves. This plan also identifies actions related to mapping, enhancing protection, educating and engaging the public, and increasing the technical capacity for exploration and stewardship. These efforts are supported by a general agreement a between the Department of Commerce (National Marine Sanctuary Program and Estuarine Reserves Division) and the Department of the Interior (Fish and Wildlife Service and National Park Service) to collaborate on efforts to improve management efficiencies, increase joint planning efforts, enhance public education, and improve law enforcement and rescue capabilities.

States have also been engaged in developing and using MPAs, with California being perhaps the most active. California authorities have been working to reach consensus on a comprehensive program of MPAs in state coastal waters under the 1999 Marine Life Protection Act (Assembly Bill 993). In April 2007, the California Department of Fish and Game adopted regulations to establish 29 marine reserves along the central California coast from Pigeon Point in San Mateo County to Point Conception in Santa Barbara County.[11] The central coast is the first of five regions to complete the planning and implementation process. The reserves include 204 square miles or approximately 18% of state waters with 85 square miles designated as no-take marine reserves. Recreational and commercial fishing interests opposed the proposal because of the reduced harvest opportunities it imposes, while environmental interests view this as a good start to more extensive protection. Other states have also taken action. For example, Hawaii enacted legislation (Act 306 in 1998) to establish a network of marine aquarium reserves along the Kona-Kohala coast of the Island of Hawaii.

Potential Benefits, Issues, and Challenges

Motivation for designating MPAs appears to be generated by the expectation that the intensity of human activities in the marine environment will continue to grow, and that this growth will exacerbate use conflicts and further degrade ecosystems. The benefit most often cited by proponents of MPAs is protection and restoration of ecosystems generally, and more specifically, valued fish populations. Additional benefits may include new educational and recreational opportunities, expanded tourism, protection of cultural resources (shipwrecks, for example), and contributions to basic science and to improved environmental conditions.

While debate about the benefits and costs of MPAs focuses most frequently on the role MPAs might play in the recovery of fish populations, they could provide other benefits.

- MPAs could be a source of baseline scientific data about current and changing conditions in the marine environment, and serve as a system so that baselines and changes could be compared among locations, particularly as climate change alters marine ecosystems. In this capacity, MPAs would provide a set of benefits not unlike one of the purposes that was articulated in legislation creating the National Estuarine Research Reserve System, a component of the federal Coastal Zone Management Program, discussed below.

- MPAs could serve as education destinations, providing opportunities for diving to observe the marine environment. Related on-shore support centers could be developed to inform a larger segment of the general public about resources protected in an MPA and the benefits that accompany the protected designation.
- MPAs could provide other environmental services, such as sequestering carbon, providing improved habitat for corals, and expanding mangroves that could dampen possible damages from hurricanes and other coastal storms.
- MPAs could protect cultural artifacts, such as shipwrecks and other places of historical significance, including places held sacred by native peoples. The first national marine sanctuary was designated to protect the site where the remains of the Civil War ironclad *Monitor* came to rest after sinking in 1862 off North Carolina.
- MPAs could become an attraction that makes them destinations for tourists and recreation activity. Activities may take place in MPAs or in nearby shore facilities, such as aquaria or museums. A unit in the National Park Service system, Buck Island Reef National Monument, with its snorkeling trail, is an example of such a place.

Commercial interests have responded that many of these benefits can be generated under current laws and programs. These interests assert that new initiatives should not duplicate other marine related legislation such as the Magnuson Stevens Fishery Conservation and Management Act (MSFCMA) or the Coastal Zone Management Act (CZMA). They also maintain that the benefits of additional protection must be weighed against the costs of constraining or prohibiting commercial activities.

Fisheries

The strongest and most vocal support of MPAs often is based on the potential role for MPAs in protecting and restoring fish populations by limiting the activities of commercial and recreational fishermen. Not only does overfishing reduce populations of desirable species, but interactions with fishing gear can result in the mortality of nontarget species (bycatch) and degrade habitat. The benefits of MPAs identified in these studies include:

- protecting individual species and biodiversity more generally within MPA boundaries;
- managing fishery populations by controlling commercial and recreational harvest rates and protecting locations where populations congregate at critical points in their life cycles, such as spawning grounds and nursery habitats;
- reducing damage to habitat;
- protecting rare, threatened, or endangered species;
- preserving or restoring the viability of representative habitats, and
- protecting portions of larger ecosystems from over-harvesting.[12]

Considerable scientific and social debate continues on the potential merits of MPAs. Some of the strongest opposition to MPAs is raised by fishing interests that could be hurt by a designation. In this debate, fishing interests question whether MPAs would have unacceptable socioeconomic consequences when compared to the benefits that would be generated. An important element in this discussion of possible tradeoffs is the recognition that MPA designations can displace resource use from protected areas to nearby areas, either transferring or raising new management issues by displacing fishing effort. MPA proponents assert that some fish are likely to stray into adjacent unprotected areas where they might benefit fishermen. However, without other management actions, protecting one area may lead to overuse or excessive harvest in other areas, moving or concentrating rather than resolving the biological management issues. Among the most contentious aspects of designating and managing MPAs for fishery management are:

- deciding whether and where MPAs might be appropriate for restoring fish populations;
- deciding whether certain fishing techniques should be limited or prohibited because they capture nontarget species and damage marine habitat;
- determining how MPA protection would be integrated with other management measures both within and outside the MPA; and
- understanding the effects of MPAs on the economic, social, and cultural well-being of nearby coastal communities.

Fishing disproportionately removes larger and older fish — often because they are more highly valued by recreational and commercial fishermen, and because of regulations protect smaller and faster-growing fish to increase stock yields. Older females produce greater numbers of eggs than the same biomass of younger females because egg production is generally proportional to fish volume. For some species, larvae produced by older females are more likely to survive because older females produce eggs with a greater amount of food reserves.[13] MPAs could be beneficial if fish remain in the protected area and grow to relatively larger sizes. Therefore, protected areas, such as MPAs, appear to be most beneficial to sedentary, reef-associated species or, more generally, to species which strongly associate with certain habitat features that cause these species to restrict their movement. In these situations, species would also benefit if unique habitat can be identified and protected.

For all the discussion of how protected areas would benefit fisheries, there remain large uncertainties about the effects of MPAs on the size and characteristics of fish populations and on ecosystem components more generally. Studies of MPAs show that population biomass, size of fish, density in a given area, and species diversity have increased within MPAs.[14] Others counter that improvements were often inevitable because of previous management failures that resulted in extremely low population levels. Some fishery experts conclude that consistent use of traditional approaches to fisheries management, such as size limits, catch limits, or seasons, are more effective management tools than establishing places where all fishing is permanently prohibited.[15]

Challenges to MPA Siting and Design

Translating the concept of MPAs into an extensive system would present multiple challenges. Many of these can be documented by tracking the experiences and issues addressed both at NOAA' s MPA Center and at other organizations around the world charged with MPA-related responsibilities, or by reviewing the types of controversies that have been raised when national marine sanctuaries have been proposed. An initial set of challenges centers on selection of sites. The approach to establishing MPAs often reflects the answers to several questions.

- What criteria — representative habitats, ecological integrity, social acceptability, degree of degradation, diversity of species, presence of endangered species, and so forth — should determine MPA locations, sizes, and boundaries?
- Should particular species or ecosystems be protected? If protection is based on ecosystems, can only portions be protected, both because it is almost impossible to protect an entire ecosystem in such a dynamic setting, and because migratory species move across many ecosystems?
- Should boundaries of an MPA be decided based on geographic areas or on ecosystems, which are often fragmented? If an MPA is adjacent to the coast, could it also include waterways that drain into it or associated terrestrial areas?
- Should sites that are already protected under other designations with purposes that are similar to MPAs be given a higher priority to receive an MPA designation, or a lower one since they are already protected?

These challenges represent only a fraction of potential siting issues. Many of the areas that are likely to be considered for MPAs already have had extensive human use, and effects of this use are frequently the reason for proposing designation. Proposals to designate units under the federal National Marine Sanctuaries Program have generated controversy because of opposition from interests who may have their activities curtailed. The designation process, which has lasted several years in some cases, has resulted in protracted debates among interests, or stakeholders, and some proposals have been rejected because of an inability to resolve these conflicts.

The success or failure of MPA designations is far more likely to reflect socioeconomic, cultural, and political factors than to reflect biological considerations. Perceptions are likely to be positive for stakeholders (including resource users), if designations are viewed as based on a fair, equitable, and transparent process for establishing and managing sites; clearly stated goals for the site; and expanding benefits while containing costs. In one example, a study compared a successful national marine sanctuary designation in American Samoa with an unsuccessful one in Puerto Rico, and attributed much of the difference in outcomes to the attention given to local cultural, social, and economic circumstances and institutions. One such difference was that success required retaining a local voice in management decisions.[16]

Marine Jurisdiction

MPAs of various sorts have been established in a range of aquatic habitats including open ocean, coastal areas, intertidal zones, estuaries, and the Great Lakes. In these areas, states and the federal government assert varying degrees of authority over activities affecting living and nonliving resources. Ownership and control also are viewed differently in marine areas, where there is little private ownership, although submerged lands, especially in nearshore areas, are sometimes leased. Generally the degree of control increases as one moves toward shore from the seaward extent of the Exclusive Economic Zone (EEZ), 200 nautical miles (nm)[17] from shore, to the baseline, a line that generally follows the shoreline. Governments also protect environmental quality, and lease the surface and subsurface for numerous activities, most notably oil and gas extraction. Zones or areas defined in ocean and coastal areas include state waters (shoreline to 3 nm), territorial sea (shoreline to 12 nm), contiguous zone (12 to 24 nm), exclusive economic zone 12 to 200 nm) and the high seas (beyond 200 nm).

The range of marine organisms and associated ecosystems seldom conform to these political and national boundaries. This presents management challenges when activities harmful to an MPA occur outside jurisdictional boundaries of the agency managing the MPA or when organisms move into other jurisdictions. Potential benefits of MPAs are likely to be dissipated unless agreements can harmonize protection efforts beyond MPA and jurisdictional boundaries

Zoning

An MPA designation might not require that an entire site be administered under one set of rules. Many proponents have advocated a zoning approach, subdividing a site into subunits with different levels of protection, or protection for different purposes. Zoning would allow managing agencies to achieve multiple policy objectives. Zoning of a protected marine area to provide different levels of protection was probably first implemented by the Great Barrier Reef Marine Park Authority in Australia in the early 1980s, and has more recently been adopted elsewhere, such as the Florida Keys National Marine Sanctuary. Subdividing designated areas may be especially useful in areas where use pressures and activities are most concentrated.

The notion of zoning is well-developed on land, but there is little experience with what adjustments might be needed to apply it effectively in an aquatic environment. One reason is that this concept is at odds with the traditional view that the ocean is "free to all" and "boundless," able to accommodate all uses in its vast expanse. This view was more widely accepted when technology to gain access to deep water resources was limited and before claims of 200-mile EEZs become an accepted international practice. National management of areas beyond the relatively narrow territorial sea is a relatively recent development and is still evolving.

In recent years, zoning of ocean areas has been getting more attention.[18] The goal of these efforts is to reduce spillover effects among competing uses that result in economic and environmental losses to society. This attention is a response to technological advances that permit greater access to and more intensive use of the marine environment, including the water column, ocean floor, and subsurface resources. At the same time, scientific research is developing a more accurate accounting of baseline conditions in the marine environment and effects of technologies on these conditions.

Other Issues

Another widely discussed issue is the desire of some to limit or prohibit offshore energy activities, including extraction and transport. This issue is highly visible, because of marine damage from oil spills around the world, often accompanied by birds or animals soaked in oil. During the 1980s and early 1990s, oil and gas development interests stressed the need to recognize multiple uses of the marine environment. They expressed concern that attempts to designate national marine sanctuaries were being used to promote a political agenda directed toward prohibiting oil and gas development in offshore areas.[19]

Several recent proposals to locate wind farms in offshore areas have raised the same kind of "not in my backyard" (NIMBY) responses that have been voiced for decades in response to unpopular development (e.g., landfills, prisons, and transmission towers) in suburban areas. Use conflicts are not new to marine areas, and similar responses have been raised to confront proposed offshore energy development in "frontier" areas where there is little prior history of such activity. More generally, thinking about future activities in the marine environment raises questions about how uses should be monitored, whether new categories of use can be effectively addressed through existing laws and programs (an issue raised by wind farm opponents, for example), and how public and private interests should be considered in any decision process.

National Reports

A number of recent national reports have studied and made recommendations concerning marine conservation and the potential use of MPAs. In particular, the 2003 Pew Oceans Commission Report and the 2004 U.S. Commission on Ocean Policy Report provide recommendations on ocean policy issues intended for policy makers. These reports, as well as the 2001 report on MPAs from the National Research Council, view the current piecemeal approach to marine resource management as contributing to the decline in marine environmental health. Many supporters of the MPA concept draw from these reports as they encourage Congress to replace the current approach with a more systematic and coordinated response. Many of these supporters have endorsed the Pew Oceans Commission and U.S. Commission on Ocean Policy recommendations as a starting place for policy discussions.

The 2001 NRC report on MPAs[20] did not make a recommendation about whether additional legislation might be desirable or whether a new law addressing MPAs is needed, although it did recognize the fragmented nature of current efforts. It discussed the costs and benefits of MPAs in comparison to more conventional management tools, explored the feasibility of implementation, and assessed the scientific basis and adequacy of techniques for the design of MPAs and marine reserves. Among its conclusions the report recommended that MPAs can be most successful if:

- all stakeholders are enlisted to participate in developing management plans;
- effective planning and design are provided; and
- integral components include regular monitoring, assessment, enforcement, and community education.

The NRC report endorsed using marine reserves as resource and fishery management tools in combination with traditional management measures. It asserted that federal and state agencies need to provide resources, expertise, and coordination for integrating individual MPAs into a framework to meet coastal and marine resource management goals established at state, regional, national, or international levels.

The Pew Oceans Commission report[21] calls for establishing a system of marine reserves as an important component of efforts to restore and maintain healthy marine ecosystems. These reserves would play critical roles in "zoning" areas of the ocean under sovereign control based on desired patterns and intensities of uses. This report also recommends establishing regional ocean ecosystems councils and a new independent federal oceans agency. The Pew Commission viewed reserves as sites that would transcend federal-state boundaries. It calls on Congress to enact a mandate to establish such a system and provide the necessary institutional structure and legal authorities to implement it. It also recommends that federal agencies should use existing authorities to establish reserves within areas that have already been designated for protection until such a mandate is enacted.

The U.S. Commission on Ocean Policy[22] also endorses MPAs, but its report is more wide-ranging than the Pew effort, and MPAs play a much smaller role in its recommendations. The report from the U.S. Commission on Ocean Policy may have the greatest impact on congressional considerations, since Congress created this commission in legislation and it reported to Congress and the Administration. The commission's recommendations include a reference to MPAs in the fisheries chapter that discusses coordinated management in federal waters and calls for a uniform process to designate, design, and monitor MPAs.[23] The commission's recommendations also include an indirect reference to MPAs related to pursuing an ecosystem approach to identify and designate "essential fish habitat" that uses current efforts to "identify important habitats and locate optimum-sized areas."

Recent Congressional Action

Currently, the National Marine Sanctuary Program is the closest to providing a comprehensive approach to using MPAs. Administrative action to establish national monuments in marine areas is gaining greater attention, but these actions appear to be opportunistic rather than deliberate movement to a comprehensive approach. Legislation proposed during the 110th Congress appears to make incremental progress in this direction, but not on the scale that most MPA advocates would prefer.

The 110th Congress has considered MPAs both directly and indirectly, but not extensively. The most direct treatment of MPAs is proposed in the Sanctuary Enhancement Act of 2008, H.R. 6537. Introduced in July 2008, H.R. 6537 would reauthorize the National Marine Sanctuary Act that was last reauthorized in 2000. Three related hearings have been held by the House Natural Resources Subcommittee on Fisheries, Wildlife, and Oceans. The proposed legislation would modify the NMSA by:

- clarifying the findings, purposes, and policies of the act and the mission of the sanctuary program;
- requiring an inventory of living and nonliving resources;
- revising sanctuary designation procedures;
- clarifying the authority to regulate fishing activities in sanctuaries;
- broadening enforcement authority of the entire system, including marine national monuments; and

- providing authority to include national monuments within the system.

Among the most important changes would be elimination of constraints on designating new sanctuaries that were included in the 2000 reauthorization, clarification of the sanctuary program mission, and integration of national monuments within the system.

Modification of previously designated national marine sanctuaries is proposed in H.R. 6204 and S. 2281, for the Thunder Bay National Marine Sanctuary and Underwater Preserve Boundary Modification Act, and H.R. 1187 and S. 2635, for the Gulf of the Farallones and Cordell Bank National Marine Sanctuaries Boundary Modification and Protection Act. The Thunder Bay NMS legislation would expand sanctuary boundaries to include additional underwater cultural resources (shipwrecks). For the Gulf of Farallones and Cordell Bank NMSs, proposed legislation would make adjustments to expand the boundaries of the NMSs. S. 2635 would prohibit certain activities within these sanctuaries, including:

- the leasing, exploration, development, production, or transporting by pipeline of minerals or hydrocarbons;
- the conduct of aquaculture (with exceptions); and
- the discharge of materials, substances, or introduced species; and
- cause a harmful change in salinity.

H.R. 1187 would also prohibit oil and gas leasing and permitting, but only regulate specific activities such as the deposit or release of introduced species and the alteration of stream and river drainage into the sanctuaries. Both bills would direct the Secretary of Commerce to complete an interim supplemental plan for each sanctuary that focuses on management in the areas added by the act and a revised comprehensive management plan for each of the sanctuaries. S. 2635 also would require a review of the Farallones NMS to determine whether the added area should be designated as a new and separate NMS.

Legislation also has been proposed to protect specific areas or to prohibit a specific activity. H.R. 5864, the Vieques Bioluminescent Bay Conservation Act, would designate Puerto Mosquito Bay as a national marine sanctuary. S. 3576 and H.R. 7051, the Georges Bank Preservation Act, would prohibit exploration, development, or production of oil or natural gas in any marine national monument or national marine sanctuary or in the fishing grounds of Georges Bank.[24]

The Coral Reef Conservation Amendments of 2007 (H.R. 1205, S. 1580, and S. 1583) are examples of proposed legislation that focus on a related topic — coral reef conservation. All three bills would identify marine protected area management strategies under the definition of conservation as a method or procedure necessary to preserve or sustain coral reefs and associated species. H.R. 1205 also would apply provisions of the National Marine Sanctuary Act to liability for costs and damages to coral reefs, even when outside national marine sanctuaries.

Congressional consideration of new and more comprehensive approaches to protect marine areas is likely to pit economic interests, who oppose MPAs because designation would place limits on the use or extraction of ocean resources, against environmental and research interests, who would like to see more widespread or systematic protection of ocean resources. MPA advocates favor resource protection over revenue-generating activities, and believe that quick congressional action is needed to prevent further destruction or deterioration of living marine resources. MPA opponents respond that the benefits of such designations are far from proven, but limitations on commercial activities are almost certain to be costly. New industries such as aquaculture and alternative energy may be especially difficult to establish if environmental interests believe they threaten marine resources. One of the main challenges for policy makers is to balance the needs for profitable private commercial activities with public concerns related to environmental quality. It is still an open question as to whether more extensive use of MPAs will assist in achieving this balance.

Appendix: Current Federal Laws and Programs

No current federal laws and programs protect marine areas as comprehensively as many of the proponents of MPAs envision for this concept. Others, especially commercial interests, counter that MPAs should be reserved for truly special areas and that current federal laws are sufficient for this purpose. Regardless, it appears that most of these laws could play significant roles if a more comprehensive effort is implemented. This discussion does not include state laws and programs, which vary widely, and would become important for MPAs that include nearshore areas or have some interaction with activities in areas under state jurisdiction.

The following laws allow designation of protected areas in the marine environment. Most apply to coastal sites, but they were enacted for different purposes and take different approaches.[25] If Congress chooses to authorize an MPA system, it might conclude that one or some combination of these programs can provide the basis for such a system, or it might conclude that it should enact entirely new authorizing legislation.

National Marine Sanctuaries Act (NMSA)

The NMSA (16 U.S.C. §§ 1431, et seq.) comes closest to authorizing what many proponents envision as MPAs. It authorizes NOAA to designate specific sites for comprehensive and coordinated management and conservation. The broad NMSA mandate allows NOAA to designate areas to preserve or restore conservation, ecological, aesthetic, or recreational values of the designated areas. It requires the development and implementation of management plans, which serve as the basis for prohibiting or limiting incompatible activities.[26]

NOAA has designated 13 sanctuaries, ranging in size from less than a square nautical mile to more than 100,000 square miles Each site was designated for a specific reason, ranging from protecting cultural artifacts to protecting entire ecosystems. At most of these sites, particularly contentious questions when developing or amending management plans have centered on which activities are incompatible with the purposes of the designation, and how incompatible activities will be limited. Since the management plans and regulations have been developed individually for each sanctuary and each sanctuary was established for a specified reason, they vary widely in how uses are managed and what uses are permitted. The

Florida Keys Sanctuary is cited as one location where the MPA concept is being applied. Within this sanctuary, 24 sites amounting to 6% of the total area have been fully protected where harvesting marine life is greatly restricted. However, this is an exception, as few of the existing sanctuaries restrict fishing, shipping, or recreation, although most prohibit oil and gas exploration and development. This has led some to characterize sanctuaries as multiple-use areas rather than areas where uses that may damage the marine environment are prohibited.

Fishing can be regulated in sanctuaries, although this has rarely occurred. In considering whether to regulate fishing, the NMSA provides the appropriate Regional Fishery Management Council the opportunity to determine whether sanctuary fishing regulations are needed and to draft fishing regulations. The Secretary of Commerce must accept the Council's proposals or determinations unless they fail to fulfill the purpose and policies of the Magnuson-Stevens Fishery Conservation and Management Act (discussed below). Regional Councils have supported only a few areas that have been closed to all fishing. As sanctuaries revise and update their management plans, whether and how to limit fishing is likely to receive more attention. In this setting, some MPA proponents do not view the sanctuary system, as it currently exists, as an effective approach for fish recovery efforts.

Coastal Zone Management Act (CZMA)

The CZMA (16 U.S.C. §§ 1451, et seq.) established a coastal zone management program and an estuarine sanctuary program, now called the National Estuarine Research Reserve System, and made 35 coastal and Great Lakes states and territories eligible to participate. The coastal zone management program provides grants to these states and territories to develop and implement plans that address several broad categories of development and resource protection activities in a state's coastal zone. Incentives to participate also include a consistency provision that requires federal actions in or affecting the coastal zone to be consistent with the state's federally approved plans. These incentives appear to be sufficient, as only one of the 35 eligible states and territories, Illinois, has not participated in the national coastal zone management program.

Places included in MPA inventories are concentrated in coastal and nearshore waters. These areas are among the most productive and diverse marine environments and are sites of the most concentrated and intensive uses and alterations. Therefore, protection efforts (including efforts to designate MPAs) also have been concentrated in these areas. Almost all coastal states have been addressing the pressures in or affecting state waters for many years, generally using their federally approved and funded coastal zone programs. A few states emphasize the marine side of their coastal zone, as management of ocean resources is one of the eight purposes for which states can receive “enhancement grants” under this program. However, most state coastal management programs generally concentrate their efforts on the land and shore side of the coastal interface.

Protecting marine areas can be addressed by states using coastal zone enhancement grants. These grants are available to participating states and territories that are successfully implementing programs and wish to do more. These grants are available for nine program areas, one of which is “planning for the use of ocean resources.” NOAA reviewed activities under this program for 1992 through 1996, and found a majority of states had some level of activity.[27] However, the organization of this study's findings makes it difficult to determine how designation and management of protected areas in state waters fit into the coastal zone management activities of these states.

The National Estuarine Research Reserve System is a component of the federal coastal zone management program. States identify research reserve sites in state waters and, after federal approval, manage them. States with research reserves have integrated them into their coastal management efforts, although the research reserves do not play identical roles. The 27 National Estuarine Research Reserves that have been designated were federally approved, in part because they each represent one of the diverse estuarine ecosystems of the marine coast (including the Great Lakes).

The system is viewed as both providing a laboratory for research and education programs, and creating a network that permits research for comparing biological or other characteristics across units of the system. Incompatible uses that would compromise the value of research reserves, such as more intensive development along the shore or major navigational improvements in the waters, are controlled or prohibited. Each research reserve operates under a management plan. The research reserves range in size from 571 acres to 365,000 acres.

Magnuson-Stevens Fishery Conservation and Management Act (MSFCMA)

The MSFCMA (16 U.S.C. §§ 1801, et seq.) established federal fishery management authority in a zone extending from the outer boundary of state coastal waters to 200 miles from the U.S. coastline. In 1976, the Fishery Conservation and Management Act[28] established eight Regional Fishery Management Councils to develop management plans for those fisheries that require active federal management. The fishery conservation zone was superceded by President Reagan's declaration of an Exclusive Economic Zone (EEZ) in March 1983 (Presidential Proclamation 5030). The EEZ applies to a broad range of resources and uses. In 1996 amendments to the act, Congress authorized the Councils to designate and manage essential fish habitat. It included the authority to regulate fishing effort up to and including closing areas to protect significant spawning and rearing habitats.[29] Such closures, which are often a response to overfishing, may be of limited duration or permanent, and they may affect some or all fishing covered by federally approved fishery management plans. Closures are usually imposed by the National Marine Fisheries Service (NMFS) on the recommendation of a Regional Fishery Management Council. Since the overarching purpose of the MSFCMA is to promote sustainable commercial and recreational use of renewable fishery resources, permanent and complete area closures remain uncommon.

Implementation of MPAs as an element of fishery management seems to be gaining interest among the Regional Fishery Management Councils. With the growing recognition that selective protection of unique habitats can benefit multiple species, Regional Councils are beginning to consider and create longer-term marine reserves (for examples, see "Recent Administrative Actions,") in lieu of temporary fishery closures for individual species or other gear or quota reductions. Partial closures, which might limit gear used, amount of fishing effort allowed, or times when fishing is allowed, are also becoming more common.

For example, the New England Regional Fishery Management Council and NMFS established closed areas on Georges Bank and adjacent areas off New England where all fishing is prohibited to foster groundfish recovery. The North Pacific Fishery Management

Council has designated a marine reserve in Southeast Alaska,[30] an Aleutian Islands Habitat Conservation Area, Aleutian Islands Coral Habitat Protection Areas, Alaska Seamount Habitat Protection Areas, a Bowers Ridge Habitat Conservation Zone, Gulf of Alaska Coral Habitat Protection Areas, and Gulf of Alaska Slope Habitat Conservation Areas.[31]

The effectiveness of fishery controls, as measured by changes in fish populations, like the effectiveness of other MPA controls, has generated controversy among competing community stakeholders. With requirements in the MSFCMA for (1) recovery schedules for overfished stocks, (2) harvest of fish at sustainable levels, and (3) minimal bycatch of fish, birds, turtles, and marine mammals, the management of U.S. marine fisheries has become significantly more restrictive since 1996. Some critics of MPAs believe that these changes in the MSFCMA reduce the need for MPAs as a fishery management tool. In 2006, the 109th Congress reauthorized the MSFCMA and included provisions to increase protection of deep sea corals as well as to provide greater emphasis on managing marine ecosystems as opposed to individual commercially valued species.

The Wilderness Act

This law (16 U.S.C. §1131, et seq.) established the National Wilderness Preservation System of congressionally designated areas of federally owned land where many activities are restricted or prohibited to minimize human alterations. Although the extension of this act's authority into marine waters is questioned by those who envision the ocean as common property, marine areas under federal jurisdiction beyond state boundaries but within 200 miles of the coastline could be eligible for designation as "wilderness" by Congress, although none has been designated to date.

Goals of the Wilderness Act are to allow unfettered operation of natural processes and provide for only those human uses, such as primitive recreational activities, that do not affect those processes. The act generally prohibits commercial activities, permanent facilities, and use of motorized equipment or motorboats, landing of aircraft, unless the use had become established before the area was designated. However, Congress has also authorized activities that do not conform with these general prohibitions. For example, the act allows for commercial uses when they are necessary "for realizing the recreational or other wilderness purposes of the area;" the use of motorboats may be authorized where such use is already established, subject to "desirable" restrictions.[32]

The prospects of establishing "marine wilderness" are being increasingly explored. Some interests who want the strongest possible protections in designated marine areas view this law as creating a model for the levels and kinds of protections that should be placed in MPAs, even if the law itself may not be readily transferred to marine areas for other reasons.[33] A portion of these interests believe that the wilderness designation should be a starting point because too many incompatible uses are still allowed. Others counter that wilderness designations are too restrictive. Many of the terrestrial wilderness debates have focused on whether the Wilderness Act's allowances for recreation, boating, or commercial use will be incompatible with protections that proponents seek through a wilderness designation.[34] If this concept is considered for marine areas, similar debates can be anticipated.

National Park Service Organic Act

This law (16 U.S.C. §§ 1, 2-4) created the National Park Service (NPS) to administer units of the National Park System, to preserve the lands and resources unimpaired, and to foster public use and enjoyment. Each unit has its own management structure; individual laws creating most of the units have placed limits on specified incompatible uses. A total of 39 NPS units in coastal areas have significant marine components. Many of these units are classified as National Seashores. Many National Park units permit recreational fishing, and a few even allow commercial fishing.[35] The Park Service's dual mandate of preservation and public use and enjoyment has resulted in conflict between interest groups who debate the desirability of providing greater access and visitor facilities versus higher levels of protection.

National Wildlife Refuge Administration Act

This law (16 U.S.C. § 668dd) establishes the primary purpose of units of the National Wildlife Refuge System to be the conservation of fish and wildlife and their habitats, and allows other compatible uses if such uses are determined to be consistent with refuge goals. The refuges are administered by the Fish and Wildlife Service in the Department of the Interior. Recreational fishing, hunting, wildlife observation, environmental education and interpretation, and nature photography are priority public uses and are allowed on many refuges; and oil and gas extraction occurs on a few units. More than 140 refuges are located along the nation's coasts, and some include offshore areas. Important functions for refuges in marine areas are managing ecosystems and providing habitat for endangered species and migratory birds as well as nursery areas that support key components of coastal and marine ecosystems. However, many regard refuge system jurisdiction as limited in the marine environment.[36]

Antiquities Act of 1906

This law (16 U.S.C. §§ 431-443) allows the President to proclaim locations of scientific or historical interest as national monuments and has been used for several marine areas.[37] Some have argued that the 1906 Antiquities Act should be used to designate protected areas in the marine environment because it can be used expeditiously. Yet it appears that applying the Antiquities Act to marine areas will still require "negotiation, education, and consensus-building" including congressional funding commitments and involvement of local committees representing interested and affected parties.[38] Use of the Antiquities Act would likely raise the same type of objections that have been voiced over other unilateral actions without the opportunity for public input and debate.[39] In the Papahānaumokuākea Marine National Monument (Northwestern Hawaiian Islands Marine Monument) case, it is likely designation was made easier because the public had already been involved during earlier consideration of the area as a national marine sanctuary.[40]

Northwest Straits (NS) Marine Conservation Initiative Act

This law (Title IV of P.L. 105-384) established the Northwest Straits Advisory Commission, and authorized the Secretary of Commerce to provide assistance to be used in accordance with the Northwest Straits Citizen's Advisory Commission Report of August 20, 1998. The NS are the waters of northern Puget Sound and southern Georgia Strait in Washington State. The priorities of the Commission are to: (1) collect marine resources data in the NS; (2) coordinate federal, state, and local marine resource protection and restoration activities in the NS; and (3) carry out other activities identified in the Report as important to such protection and restoration. Under this authority, seven county Marine Resource Committees are advising the Commission in carrying out these priorities. The 2004 program evaluation found that the initiative has generated local support of projects and conservation, increased voluntary compliance with conservation goals, brought people together to work cooperatively and exchange innovative ideas on issues, and created a model of marine governance that may be adapted to other regions.[41]

Other Protection Efforts

Many other federal laws affect the quality of the marine environment by regulating coastal and offshore activities. These laws typically set minimum environmental quality standards or protect certain elements of the marine environment rather than designate areas for use or protection. Particularly noteworthy laws in this group include the Endangered Species Act, the Clean Water Act, and the Marine Mammal Protection Act.

Certain offshore areas in federal waters were protected specifically from oil and gas development activities for over two decades. Starting with the FY1982 Interior appropriations act (P.L. 97-100), Congress annually prohibited these activities in certain areas, including waters off New England, the Mid-Atlantic states, portions of Alaska and California, the Pacific Northwest, and the Eastern Gulf of Mexico. In 1990, President Bush issued a directive limiting Outer Continental Shelf (OCS) activities to federal waters off Texas, Louisiana, Alabama, and portions of Alaska. In 1998, President Clinton extended this moratorium to OCS activities in other areas through 2012. The 109th Congress enacted the Gulf of Mexico Energy Security Act of 2006 (Division C, Title I, of P.L. 109-432), which opened up a portion of the Eastern Gulf of Mexico that had previously been closed. On July 14, 2008, President Bush lifted the executive ban on OCS activities. Congress then allowed the ban on drilling in areas of the OCS to expire by not including the moratorium in the Consolidated Security, Disaster Assistance, and Continuing Appropriations Act, 2009 (P.L. 110-329) that was signed by the President on September 30, 2008.[42]

In addition to the federal laws with authority over MPA designation and management, state and local laws as well as numerous international agreements and conventions have marine protection components. According to the MPA Center inventory, nearly 1,300 MPAs are managed by states and territories.

END NOTES

[1] International Union for the Conservation of Nature and Natural Resources (IUCN), *Guidelines for Protected Area Management Categories*, Gland, Switzerland, 1994.

[2] Although many of these areas have been in existence for decades, in 2007 the MPA Center completed its inventory of MPAs in the United States that was based on its MPA definition.

[3] For example, less than 3% of total MPA area in the United States is no-take (prohibits fishing) or less than 1% of all U.S. waters.

[4] The first plan, *Turning to the Sea: America's Ocean Future*, was announced by a Cabinet- level task force created at the National Ocean Conference in 1998. It was released September 2, 1999. The second was the *National Action Plan to Preserve Coral Reefs*, prepared by the U.S. Coral Reef Task Force and released on March 2, 2000.

[5] The order also directed the EPA to reduce pollution of beaches, coasts, and oceans by strengthening water quality protection for marine waters.

[6] A request for nominations was published at 65 *Fed. Reg.* 50503 (August 18, 2000).

[7] "Presidential Amendment Ensures Recreational Fishing to Be Maintained on Federal Lands and Waters," *PR Newswire* (Sept. 27, 2008).

[8] For information on the activities of the National MPA Center, go to [http://www.mpa.gov].

[9] The most recent revised draft framework is available at [http://mpa.gov/national system /national system.html].

[10] The inventory is available at [http://mpa.gov/helpful_resources/inventory.html].

[11] For information related to the California Marine Life Protection Act Initiative, see [http://www.dfg.ca.gov/mlpa/phase1.asp].

[12] Stephan R. Palumbi, Marine Reserves: *A Tool for Ecosystem Management and Conservation*, prepared for the Pew Oceans Commission (Palo Alto, CA: Stanford Univ., 2002), 44 pp.

[13] Stephen R. Palumbi, "Fisheries Science: Why Mothers Matter," *Nature,* v. 430, (August 5, 2004), pp. 621-622.

[14] PISCO (Partnership for Interdisciplinary Studies of Coastal Oceans), *The Science of Marine Reserves,* 2007, at [http://www.piscoweb.org].

[15] Dr. Robert L. Shipp, *Take Marine Protected Areas (MPAs) as a Fishing Management Tool: A Programmatic Perspective*, Report to the FishAmerica Foundation, (Alexandria, VA: 2002).

[16] S. J. Fiske, "Sociocultural Aspects of Establishing Marine Protected Areas," *Ocean and Coastal Management*, v. 17, no. 1 (1992): 25-46.

[17] A nautical mile (nm) (6,076 feet) is equal to 1.15 statute miles (5,280 feet).

[18] Garry Russ and Dirk Zeller, "*From Mare Liberum to Mare Reservarum,*" *Marine Policy*, v. 27 (2003): 75-78.

[19] Statement of William P DuBose, Vice President, National Ocean Industries Association, Hearing on the Current Status and Future Needs of the National Oceanic and Atmospheric Administration's National Marine Sanctuary Program, before the House Merchant Marine and Fisheries Subcommittees on Oceanography, Great Lakes, and Outer Continental Shelf Subcommittee and on Fisheries and Wildlife Conservation and the Environment, (Nov. 7, 1991).

[20] National Research Council (NRC), M*arine Protected Areas: Tools for Sustaining Ocean Ecosystems, National Academy Press* (Washington DC:2001).

[21] Pew Oceans Commission, *America's Living Oceans: Charting a Course for Sea Change*, The Pew Charitable Trusts, Pew Environmental Group (Philadelphia, PA:2003). The Pew Oceans Commission website and links to related reports may be accessed at [http://www.pewtrusts.org/our work detail.aspx?id=130].

[22] U.S. Commission on Ocean Policy, *An Ocean Blueprint for the 21st Century,* Final Report, (Washington DC:2004).

[23] Recommendations 19-21 and 6-3, respectively. The Oceans Commission website and links to the final report may be accessed at [http://www.oceancommission.gov/].

[24] Georges Bank, located off the coast of New England, is one of the richest fishing grounds in the United States.

[25] For a more detailed review of current federal programs and laws, see National Research Council, "Chapter 8: Historical Background and Evaluation of *Marine Protected Areas in the United States," Marine Protected Areas Tools for Sustaining Ocean Ecosystems* (Washington DC: National Academy Press, 2001), pp. 145-173.

[26] For a detailed review of the legislative history of the NMSA, see William L Chandler and Hannah Gillelan, "The History and Evolution of the National Marine Sanctuaries Act," *Environmental Law Reporter*, no. 34 (2004): 10506-10565.

[27] U.S. Dept. of Commerce, NOAA, Office of Ocean and Coastal Resource Management, Coastal Programs Division, *State Enhancement Grant Assessments and Strategies: Ocean Governance*, NOS/OCRM/CPD 990-08,(Washington, DC: Oct. 1999), 63 p.

[28] The name was changed to the MFCMA and later to the MSFCMA in recognition of the contributions of Senators Warren Magnuson and Ted Stevens to federal fisheries management.

[29] Similar closures occur in state coastal waters under the authority of various state laws and interstate compacts as well as internationally under the authority of negotiated conventions and agreements.

[30] 65 *Fed. Reg*. 67305 (November 9, 2000).

[31] 71 *Fed. Reg*. 36694 (June 28, 2006).

[32] See CRS Report RL33827, *Wilderness Laws: Permitted and Prohibited Uses*, by Ross W. Gorte.

[33] Marine Wilderness Committee, Report on the Concept of Marine Wilderness, AFS Policy Statement #18, See [http://www.fisheries.org/afs/docs/policy_18f.pdf].

[34] Marine Sciences Research Center, *First Regional Marine Wilderness Areas Workshop*, Stony Brook University, (Stony Brook, NY: Nov. 2001), p.40. See [http://www.somas.st onybrook.edu/news/MWW1/proceedings.pdf]

[35] In one of these areas, Glacier Bay National Park and Preserve, Alaska, commercial fishing is currently being phased-out.

[36] A Department of Justice report of September 15, 2000, by Randolph D. Moss, Assistant Attorney General, concluded: "We are unconvinced, however, that the President would have the authority to establish a national wildlife refuge in either the territorial sea or the EEZ...."

[37] 16 U.S.C. §§ 431-433.

[38] Jeff Brax, "Zoning the Ocean: Using the National Marine Sanctuaries Act and the Antiquities Act to Establish Marine Protection Areas and Marine Reserves in America," *Ecology Law Quarterly*, v. 29 (2002): 71-129.

[39] See CRS Report RS20902, *National Monument Issues*, by Carol Hardy Vincent.

[40] The proclamation establishing the Papandnaumokudkea Marine National Monument can be found at [http://www.whitehouse.gov/news/releases/2006/06/print/20060615-18.ht ml].

[41] See [http://www.nwstraits.org/Page1D/181/default.aspx].

[42] See CRS Report RL33493, *Outer Continental Shelf: Debate Over Oil and Gas Leasing and Revenue Sharing,* by Mark Humphries.

In: Marine Protected Areas
Editor: Floyd B. Mayr

ISBN: 978-1-60741-549-7

Chapter 2

MARINE PROTECTED AREAS (MPAS): FEDERAL LEGAL AUTHORITY

Adam Vann

SUMMARY

Recent events, including the release of the President's U.S. Ocean Action Plan and reports issued by the United States Commission on Ocean Policy and the Pew Oceans Commission, have prompted a reexamination of U.S. ocean policy and debate over an "ecosystem approach" to ocean resource management. One proposed mechanism for conserving ocean resources is the Marine Protected Area (MPA), conceptualized as a zoning system for the portions of the ocean under U.S. jurisdiction. This has been highlighted by the issuance of the Revised Draft Framework for Developing the National System of Marine Protected Areas (Draft Framework), issued on March 17, 2008, by the National Marine Protected Areas Center, part of the National Oceanic and Atmospheric Administration (NOAA). The introduction of H.R. 21 in the 110th Congress has also focused attention on the health of ocean resources.

The relative merits and the potentially negative consequences of such an MPA system have been widely discussed. Advocates of additional protection argue that a more comprehensive system as outlined in the Draft Framework should be established. Others argue that the current system is effectively managing ocean resources and that additional restrictions would be economically harmful.

Apart from the relative merits of each position, there is some question as to the applicability of current federal law to the oceans and whether new protections could be imposed administratively, without additional legislation. To some extent, regulatory authority depends upon the nature of the jurisdiction that the United States has claimed over various ocean resources vis-a-vis other nations and vis-a-vis the states. Consistent with international law, the United States claims jurisdiction over marine areas extending 200 nautical miles from its coast and has regulated resources in the zones composing this area under multiple legal authorities.

Several current laws which might provide authority for the creation of MPAs are aimed specifically at the ocean environment. The National Marine Sanctuary Program, established by the Marine Protection, Research and Sanctuaries Act, the Magnuson-Stevens Fishery Conservation and Management Act, and the Coastal Zone Management Act each specifically contemplate various levels and forms of aquatic resource protection.

Additionally, certain generally applicable laws, while primarily intended for use on land, would arguably support the designation of an MPA in some circumstances. Indeed, U.S. MPAs within the territorial seas have been established as national monuments, national parks, national wildlife areas, and, most recently, as a reserve via executive order.

This report outlines U.S. jurisdiction over ocean resources and analyzes the existing laws to assess their application to marine environments.

INTRODUCTION

Scientific documentation of the ecological condition of the oceans, such as the reports recently issued by the United States Commission on Ocean Policy and the Pew Oceans Commission, has been cited as evidence of deteriorating aquatic conditions. The President responded to these reports by creating the Committee on Ocean Policy within the White House Council on Environmental Quality.[1] The Bush Administration has also released a U.S. Ocean Action Plan, a report setting forth Administration support for development of a Global Ocean Observing System; various state, local, and federal partnerships in ocean stewardship and management; individual fishing quotas and greater use of market-based systems for fisheries management; development of an Ocean Research Priorities Plan and Implementation Strategy; and a National Freight Action Agenda to ensure a safe, reliable, efficient, and competitive freight transportation system. The Plan also indicates that the President will pursue Coral Reef Local Action Strategies, U.S. accession to the UN Convention on the Law of the Sea, and passage of legislation to more clearly define the National Oceanic and Atmospheric Administration's (NOAA's) responsibilities within the Department of Commerce.[2]

Recently there has been debate over whether the development of a more comprehensive system of marine resource preservation is appropriate.[3] Currently, a number of U.S. marine sites have been designated for and receive special protections under laws specifically aimed at preserving ocean resources, such as the National Marine Sanctuary Program, established in 1972 by the Marine Protection, Research and Sanctuaries Act;[4] the Magnuson-Stevens Fishery Conservation and Management Act;[5] and the Coastal Zone Management Act.[6]

Executive Order 13,158, issued by President Clinton and retained by the Bush Administration, made Marine Protected Area (MPA) designation and management a national priority. That Order defined an MPA as "Any area of the marine environment that has been reserved by federal, state, territorial, tribal or local laws or regulations to provide lasting protection to part or all of the natural or cultural resources therein."[7] Under Executive Order 13,158, federal agencies are directed to strengthen general protections for existing MPAs and to prevent federal actions from resulting in harm to these areas. Agencies are also directed to improve management efforts, in part through establishing a comprehensive national MPA system.[8] The Executive Order does not clarify where (i.e. which zone of U.S. jurisdiction)

MPAs can be created or what laws authorize their creation. It should be noted, however, that Executive Order 13,158 does not confer new designation or management authority on the federal agencies, stating that "[e]ach Federal agency *whose authorities provide for* the establishment or management of MPAs shall take appropriate actions to enhance or expand existing MPAs and establish or recommend, as appropriate, new MPAs."[9] Additionally, the Order states that when designating MPAs, federal agencies must "act in accordance with international law and with Presidential Proclamation 5928 of December 27, 1988, on the Territorial Sea of the United State of America, Presidential Proclamation 5030 of March 10, 1983, on the Exclusive Economic Zone of the United States of America, and Presidential Proclamation 7219 of September 2, 1999, on the Contiguous Zone of the United States."[10]

In September of 2006, the National Marine Protected Areas Center, a division of the National Oceanic and Atmospheric Administration, issued its Draft Framework for Developing the National System of Marine Protected Areas.[11] After an extensive comment period, the National Marine Protected Areas Center produced a revised version of this document in March of 2008 (referred to herein as the Draft Framework). The Draft Framework provides guidance for a collaborative effort among federal, state, tribal, and local governments and other stakeholders to develop an effective National System of Marine Protected Areas, building on existing programs.[12] However, as the Draft Framework acknowledges, "[n]either the [MPA] national system nor [Executive Order 13,158] establish any new legal authorities to designate or manage MPAs, nor do they alter any existing federal, state, local, or tribal MPA laws or programs. Each MPA or program that participates in the national system will continue to be independently managed by its respective agency or agencies, as will any new sites that eventually may be established."[13]

As the Draft Framework states, the new National System is intended to be a "system of sites and systems" coordinating the various existing programs with legal authority to establish MPAs.[14] This report analyzes various sources of legal authority to assess their possible application to marine environments and will outline the protection and management system each might support.

U.S. Jurisdiction over Marine Resources

International law recognizes that coastal nations have legal authority to manage certain ocean resources within their jurisdiction. The 1982 United Nations Convention on the Law of the Sea (UNCLOS III) recognizes general zones within which signatory nations may regulate exploitation of marine resources. UNCLOS III recognizes a region extending up to twelve "nautical miles"[15] from a nation's coast, the territorial sea of that nation, in which the coastal nation may claim full ownership and sovereignty over the waters, seabed, and the subsoil.[16] Coastal nations can further regulate beyond the territorial sea up to 24 nautical miles from the coast, the contiguous zone, in so far as necessary to protect the territorial sea and to enforce its customs, fiscal, immigration, and sanitary laws.[17] Further, UNCLOS III allows for an exclusive economic zone (EEZ), which extends two hundred nautical miles from the coast. In its EEZ, the coastal nation has sovereign rights to explore, exploit, conserve, and manage marine resources.[18] Many elements of the jurisdictional scheme under UNCLOS III reflect long-standing practice and may be considered customary international law, a position the United States appears to have taken when it proclaimed its own EEZ jurisdiction.[19]

Although the United States has signed the most current version of the UNCLOS III agreement, it has yet to ratify the treaty; consequently, the United States is not a formal party. Even absent ratification, however, the U.S. has claimed jurisdiction over zones virtually identical to those contemplated by UNCLOS III via a series of presidential proclamations.[20]

Several federal laws explicitly apply to U.S. waters, including the territorial sea, contiguous zone, and EEZ.[21] Additionally, certain resource protection and management frameworks now in place might also be applied to ocean resources in the territorial sea, the contiguous zone, and the EEZ should Congress or the President designate such areas for protection. Generally, Congress has broad constitutional authority to "dispose of and make all needful Rules and Regulations respecting the Territory or other Property belonging to the United States...."[22] Also, any portions of the U.S.-claimed waters and resources that might not be fairly characterized as "Territory" or "Property" might nonetheless fall under congressional authority to "regulate Commerce with foreign Nations and among the several States, and with the Indian Tribes...." among others.[23]

The President, even if not explicitly granted regulatory authority by statute, may also have constitutional authority to impose regulations on the use of ocean resources claimed by the United States. For instance, the Constitution grants the President broad authorities regarding the foreign relations, national defense, and treaties of the United States.[24]

Additionally, should UNCLOS III be ratified, its many marine habitat provisions, which require parties to protect and preserve the marine environment,[25] may provide the President and Congress with additional authority for the regulation of marine resources in order to execute the treaty obligations of the U.S. The President and Congress may also have authority to protect ocean resources under the auspices of implementing current U.S. treaty obligations. Under the Protocol Amending the 1916 Convention for the Protection of Migratory Birds, for instance, the government of a signatory nation is directed to "use its authority to protect and conserve habitats essential to migratory bird populations."[26] This might authorize the protection of ocean resources to the extent necessary to implement the purpose of the underlying treaty.

The current extent of the authority of the U.S. government in each zone is complex and must be considered in two contexts: the federal government vis-a-vis the international community and the federal government vis-a-vis the states. An overview of the relative authorities in each zone follows.

The Territorial Sea

As stated above, UNCLOS III recognizes a territorial sea extending twelve nautical miles from a nation's coast in which a coastal state may exercise full jurisdiction to the extent that it does not conflict with the right of foreign vessels to innocent passage.[27] The United States claims a twelve nautical mile territorial sea consistent with the UNCLOS III expression. Presidential Proclamation 5928 states that the United States "exercises sovereignty and jurisdiction ... that extend to the airspace over the territorial sea, as well as to its bed and subsoil."[28] Thus, the United States is generally considered to exercise full sovereign authority over its territorial sea vis-a-vis other nations. The United States Supreme Court has also recognized U.S. authority to impose significant protective measures on ocean resources in this area.[29] Thus, it would appear relatively clear that, in the international law context, U.S. jurisdiction over the territorial sea is analogous to the sovereignty a nation possesses over its land territory, subject to the right of innocent passage.[30]

Jurisdiction over the territorial sea of the United States is complicated by the authority of coastal states under our federal system. The Federal Submerged Lands Act of 1953[31] assured coastal states title to the lands beneath coastal waters in an area stretching, in general, three "geographical miles"[32] from the shore.[33] Thus states may regulate the coastal waters within this area, subject to federal regulation for "commerce, navigation, national defense, and international affairs ..."[34] and the power of the federal government to preempt state law.[35] The remaining outer portions of waters over which the United States exercises jurisdiction are federal waters.[36]

The Exclusive Economic Zone

Consistent with UNCLOS III and international law and custom, the United States has claimed an EEZ extending, in general, 200 nautical miles from its coasts.[37] In its EEZ, the United States has sovereign rights over the exploration, exploitation, conservation, and management of the natural resources of the sea-bed, subsoil, and the superadjacent waters.[38] According to UNCLOS III, U.S. jurisdiction also extends over "other activities for the economic exploitation and exploration of the zone, such as the production of energy from the water, currents and winds"[39] and, subject to some limitations, "the establishment and use of artificial islands, installations and structures; marine scientific research; and the protection and preservation of the marine environment."[40] While the United States does claim sovereign rights over natural resource management and the regulation of certain economic activities, it has less than full sovereignty and ownership of its EEZ.[41] As with the contiguous zone, it remains unclear precisely how this might limit U.S. regulation or the application of federal law currently aimed at areas "within the United States."

The Contiguous Zone

Under UNCLOS III and customary international law, a coastal nation may claim a contiguous zone extending beyond its territorial sea and up to twenty-four nautical miles from the coast, in which a coastal nation may also claim jurisdiction and regulate as may be necessary to protect the territorial sea and to enforce its customs, fiscal, immigration, and sanitary laws.[42] After the extension of the U.S. territorial sea, President Clinton issued Proclamation No. 7219, claiming a U.S. contiguous zone reaching twenty-four nautical miles from the coast.[43] Consistent with UNCLOS III, the Proclamation states that the United States "may exercise the control necessary to prevent infringement of its customs, fiscal, immigration, or sanitary laws and regulations within its territory or territorial sea."[44]

The exact contours of U.S. authority in the contiguous zone are not, however, clearly defined. In *United States v. De Leon*, the Court of Appeals for the First Circuit stated "[t]he contiguous zone is an area in which the United States claims certain rights short of sovereignty."[45] The court did not go on to define the U.S. rights in the area nor does it appear that other courts have had the opportunity to do so. Thus, while it is clear that Congress has directed legislation at contiguous zone resources and that the United States can exercise some amount of regulatory control over the zone, it would not appear that this area would constitute U.S. territory subject to full U.S. sovereignty or ownership.[46] Accordingly, the regulatory authority of the United States or the applicability of federal laws directed at areas "within the United States" may be limited.

Current Law Specific to Marine Environments

National Marine Sanctuaries Act

The National Marine Sanctuaries Act (NMSA),[47] found in Title III of the Marine Protection, Research, and Sanctuaries Act of 1972, comes the closest to providing the legal framework for the creation of a national system of marine protected areas.[48] The scope of this act looks beyond species or resource-specific protection and focuses on protecting entire marine ecosystems (an approach also adopted under the essential fish habitat provisions of the Magnuson-Stevens Fishery and Conservation Management Act, discussed *infra*).[49] Thirteen national marine sanctuaries, covering approximately 18,000 square miles in the Atlantic and Pacific Oceans, have been established.[50] While these sites vary significantly as to geographic region and the types of resources they contain, it would appear that each is located in the territorial sea of the United States or one of its island territories.[51]

Designation of national marine sanctuaries under the act can take place by act of Congress or administratively.[52] As to administrative designation, the Secretary of Commerce (the Secretary), acting through NOAA, is authorized to designate "any discrete area of the marine environment as a national marine sanctuary and promulgate regulations implementing the designation...."[53] Administrative designation under the act requires compliance with a statutorily imposed process, described below.

The administrative designation process begins with publication of several documents in the *Federal Register*, including a notice of the proposal.[54] The Secretary must also furnish the "terms of the proposed designation,"[55] the draft management plan, proposed regulations, the draft Environmental Impact Statement (EIS), and cost estimates, as well as other supporting documents.[56] The law requires a public hearing to be held in the coastal area or areas most affected by the designation.[57] The act provides for congressional oversight of the designation process, allowing the House Committee on Resources and the Senate Committee on Commerce, Science, and Transportation to issue reports that the Secretary must consider before publishing notice of an intent to designate.[58] Further, no designation may be proposed unless the Secretary finds the new designation will not have a negative impact on the existing system and there are sufficient fiscal resources for effectively implementing the management plan and complying with various site study requirements. Having complied with these procedural requirements, the Secretary may designate a marine sanctuary upon finding that:

(1) the designation will fulfill the purposes and policies of [the NMSA];
(2) the area is of special national significance due to

(a) its conservation, recreational, ecological, historical, scientific, cultural, archaeological, educational, or esthetic qualities;
(b) the communities of living marine resources it harbors; or
(c) its resource or human-use values;

(3) existing State and Federal authorities are inadequate or should be supplemented to ensure coordinated and comprehensive conservation and management of the area, including resource protection, scientific research, and public education;
(4) designation of the area as a national marine sanctuary will facilitate the objectives stated in paragraph (3); and
(5) the area is of a size and nature that will permit comprehensive and coordinated conservation and management.[59]

The act also sets forth a list of factors to consider in making the above findings and requires agency consultation with interested state and federal authorities as well as other interested persons.[60] A final designation does not take effect until notice of the designation decision, the availability of a final EIS and management plan, and publication of the final regulations implementing the plan.[61] Designations within the seaward boundaries of a state are also subject to approval by the Governor.[62]

Regulation of marine sanctuaries can vary significantly from site to site. The NMSA does not prescribe specific protections for sites designated under its authority and, in fact, encourages multiple uses.[63] Thus, unlike national parks, which generally receive stringent ecological protection, sanctuaries designated under this act frequently allow fishing and shipping activities.[64]

As part of the EIS for a proposed sanctuary, the Secretary must prepare and publish a resource assessment report documenting "present and potential uses of the area," with an emphasis on compatible uses such as fishing, energy development, research, and recreational uses.[65] Further, the Secretary must consider "the negative impacts produced by management restrictions on income-generating activities such as living and nonliving resources development" and "the socioeconomic effects of sanctuary designation."[66] Fishing interests are afforded special protection. The Secretary is directed to consider present commercial and recreational fishing interests when making a designation.[67] The appropriate Regional Fishery Management Council[68] is given the first opportunity to draft all fishing regulations "as the Council may deem necessary to implement the proposed designation!"[69] The Secretary is directed to accept the Council's proposed rules "unless the Secretary finds that the Council's action fails to fulfill the purposes and policies of this chapter and the goals and objectives of the proposed designation."[70] Further, a designation will not terminate "a valid lease, permit, license, or right of subsistence use or access in existence on the date of designation!"[71] The Secretary is empowered to regulate the exercise of such rights; however, fishing is expressly excluded from the activities requiring a special-use permit under the act.[72]

While the act does not provide for specific protections, it does authorize the Secretary to undertake "all necessary actions" to prevent or respond to damage to a marine sanctuary.[73] Such actions are funded, at least in part, by any damages received from the party responsible for a particular injury.[74] Thus, for instance, when illegal poaching or an oil spill takes place, resources may be rehabilitated with recovered funds. Liability under the act has been interpreted broadly, with the Eleventh Circuit holding that the NMSA imposes strict liability for injuries to protected marine resources.[75]

Magnuson-Stevens Fishery Conservation and Management Act

The Magnuson-Stevens Fishery Conservation and Management Act (Magnuson-Stevens)[76] establishes a "national program for the conservation and management of the fishery resources of the United States ... to prevent overfishing, to rebuild overfished stocks, to insure conservation, and to realize the full potential of the Nation's fishery resources."[77] This law establishes a regulatory system applicable to management of domestic fisheries within U.S. waters, excluding the region coastal states control under the Federal Submerged Lands Act.[78]

Magnuson-Stevens gives primary responsibility for the nation's marine resources to eight regional Fishery Management Councils.[79] For fisheries within their region, the Councils prepare and implement Fishery Management Plans (FMPs), any subsequent FMP amendments, and fishery regulations, all subject to prescribed national standards.[80] The crux of this act directs that fishery management, through the above-mentioned FMPs and FMP implementing regulations, "prevent overfishing while achieving, on a continuing basis, the optimum yield from each fishery for the United States fishing industry."[81] "Optimum yield" is defined as the amount of fish which:

> (A) will provide the greatest overall benefit to the Nation, particularly with reference to food production and recreational opportunities, and taking into account the protection of marine ecosystems;
> (B) is prescribed on the basis of the maximum sustainable yield from the fishery, as reduced by any relevant social, economic, or ecological factor; and
> (C) in the case of an overfished fishery, provides for rebuilding to a level consistent with producing the maximum sustainable yield in such fishery.[82]

The act also defines "overfished" as the rate at which mortality "jeopardizes the capacity of a fishery to produce the maximum sustainable yield on a continuing basis."[83]

The regulations and FMPs submitted by the regional councils are reviewed by the Secretary of Commerce and are subject to the Secretary's approval.[84] While Magnuson-Stevens establishes a basic policy and framework for fishery regulation and management, it does not impose specific, blanket requirements for such activities. Indeed, it suggests possible courses of action, including a broad provision authorizing such "measures, requirements, or conditions and restrictions as are determined to be necessary and appropriate for the conservation and management of the fishery" in order to accommodate the needs of divergent sites.[85] Typical management techniques include limiting access for seasonal or more indefinite terms, applying fishing quotas (often referred to as Total Allowable Catch), imposing gear restrictions (such as regulating the mesh size used in nets to control the size of fish and/or the species taken), and taxing the amount of fish caught.[86] Magnuson-Stevens authorizes additional preservation authorities, empowering regional councils to ensure compliance with FMPs and regional regulations. The act authorizes regional councils to close fisheries to all exploitation in order to remedy or prevent overfishing, as defined by the act and its regulations.[87] Regionally applicable regulations and rules geared toward specialized situations provide more specific guidance as to the procedures for fishery closure.[88]

Additionally, Magnuson-Stevens requires that regional councils, through FMPs, protect "essential fish habitat" (EFH), an authority that is, in some respects, comparable to the ability to designate MPAs.[89] The law requires regional councils to identify EFH for species in need of protection and, with the aid of an ecosystem panel, to develop a plan to conserve and enhance EFH.[90] The National Marine Fishery Service has further refined the EFH statutory requirements in its regulations. Under these regulations, FMPs must identify those species in need of protection, and then identify and designate the EFH for that species.[91] While focusing on protecting particular species, the regulations adopt an "ecosystem approach" to defining EFH, stating "EFH should be based on ... the quality and quantity of habitat that is necessary to maintain a sustainable fishery and the managed species' contribution to a healthy ecosystem.... where ecological productive capacity is maintained, diversity of the flora and fauna is preserved, and the ecosystem retains the ability to regulate itself."[92] Once EFH has been designated, regional councils are required to "prevent, mitigate, or minimize" degradation of the EFH resulting from fishing operations.[93] As is generally true under Magnuson-Stevens, no specific protection or mitigation measures are required, leaving open a variety of protection options.[94]

Management flexibility is a key goal under Magnuson-Stevens, with the act and its implementing regulations recognizing the need for adaptive regulatory techniques. The act requires periodic review and continuous assessment of FMPs and allows for necessary amendments.[95] FMP implementing regulations may also be amended "on a timely basis, as new information indicates the necessity for change in objectives or management measures."[96] In addition, the regulations also take into account the need for flexibility in fishery management, allowing for certain, necessary modifications of regulatory techniques as conditions in each fishery require. To this end, the general regulations issued under Magnuson-Stevens specifically recognize the uncertainties inherent in the planning process and encourage FMPs to include multiple regulatory options that can be implemented as needed without amending FMPs or their own implementing regulations.[97]

Thus, Magnuson-Stevens, while aimed at sustaining fish stocks and providing broader ecosystem protection, establishes a framework which can provide for flexible and responsive management of marine resources. The authority to set optimum yield, catch quotas, and to close fisheries as necessary to protect fish populations provides regulators with significant tools for implementing fishery preservation.

Outer Continental Shelf Lands Act

The Outer Continental Shelf Lands Act of 1953, as amended, (OCSLA)[98] establishes exclusive federal jurisdiction over all submerged lands lying seaward and outside of the areas designated by the Submerged Lands Act as under state jurisdiction.[99] The OCSLA authorizes the Secretary of the Interior to grant mineral leases on Outer Continental Shelf (OCS) lands "to the highest responsible qualified bidder or bidders by competitive bidding."[100] While providing for orderly development of OCS mineral resources, the OCSLA provides the Secretary of the Interior with broad leeway to refrain from offering areas for mineral development.

The Secretary of the Interior is instructed to prepare a comprehensive "oil and gas leasing program."[101] This program is required to provide for a five-year leasing schedule, documenting the size, timing, and location of foreseeable leasing activity and must consider "economic, social, and environmental values of the renewable and nonrenewable resources contained in the outer Continental Shelf, and the potential impact of oil and gas exploration on other resource values of the outer Continental Shelf and the marine, coastal, and human environments."[102]

The Secretary of the Interior has broad authority to impose conditions on the development of OCS resources governed by the OCSLA and may refrain from leasing areas for development altogether.[103] When a lease has been approved, a lessee must submit development plans and permit applications to the Secretary of the Interior at each development stage.[104] The Secretary of the Interior is directed to disapprove a particular plan or cancel a lease should the Secretary find that such plan or lease will necessarily result in serious harm to environmental or mineral resources.[105] In addition to whatever environmental or resource impacts the Secretary of the Interior may consider in reviewing a development plan, development plans must be consistent with state regulation as provided for under the Coastal Zone Management Act (discussed below).[106] States may prevent Interior approval of a development plan unless the Secretary of Commerce overrides a state finding of inconsistency with coastal zone management regulations.[107]

Thus, the OCSLA provides authority for ensuring environmentally sensitive mineral development. Further, in conjunction with other resource-oriented statutes, such as the Magnuson-Stevens Fishery Conservation and Management Act and state coastal zone management plans, statutory authority to withhold areas from production[108] administratively could provide significant preservation authority.

Coastal Zone Management Act

As discussed above, the jurisdiction of coastal states, in most cases, extends three geographical miles from the shoreline under the Federal Submerged Lands Act, giving states primary regulatory responsibility for preservation and regulation of the nation's coastal areas.[109] The Coastal Zone Management Act (CZMA)[110] was designed to encourage states to enact coastal zone management plans to coordinate protection of habitats and resources in coastal waters. The act establishes a policy of preservation alongside sustainable use and development when such activities are compatible with resource protection.[111]

Programs under the CZMA are managed by the states, the Department of Commerce through NOAA, and the Environmental Protection Agency (EPA). Under the act, state coastal zone management programs that are approved by the Secretary receive federal monetary and technical assistance. To qualify for federal funds state programs must designate land and water conservation measures and permissible uses,[112] and must address various sources of water pollution.[113] The CZMA also requires that the federal government and federally permitted activities comply with state programs.[114]

GENERAL PRESERVATION LAWS

National Monuments

The Antiquities Act[115] has been used to designate many national monuments since the law's enactment in 1906 and, on several occasions, has been the basis for setting aside marine areas for protected status. The Act delegates a broad authority to the executive branch and, in relevant part, states:

> The President of the United States is hereby authorized, in his discretion, to declare by public proclamation historic landmarks, historic and prehistoric structures, and other objects of historic or scientific interest that are situated *upon the lands owned or controlled by the Government of the United States* to be national monuments, and may reserve as a part thereof parcels of land, the limits of which in all cases shall be confined to the smallest area compatible with the proper care and management of the objects to be protected.[116]

While the precise language of the statute might appear to limit monuments to "landmarks," "structures," or other objects which are "situated on the lands" and limits what may be reserved to "parcels of land," the Executive's power under the Antiquities Act has been interpreted quite broadly:[117] The act has been interpreted to cover submerged lands under U.S. jurisdiction. In fact, submerged lands (e.g., coral reefs) have been included in national monuments on several occasions, but always, apparently, in connection with protection of some associated surface lands, such as islands.[118] In addition to the submerged lands themselves, the act has been "read to authorize protection of the water column above submerged lands as well," provided that a qualifying object located on the submerged lands could not be adequately protected absent water column protection.[119] This reading of the act was apparently embraced by the Clinton Administration when the President expanded protection for the Buck Island Reef National Monument, where, in addition to protection for submerged lands themselves, fishing is now prohibited.[120] A monument designation protecting only water resources, and not an underlying land- based monument, would seem to strain the language of the statute. It should be noted again, however, that the act has been interpreted broadly, and it does not appear that there has ever been a successful challenge to a designation.

An additional issue is whether the U.S. government "owns" or "controls" these submerged lands as contemplated by the Antiquities Act. This question is complicated by the breakdown of the sea into various jurisdictional zones. Ownership and control may vary depending on which seaward area is involved. As discussed above, the United States asserts full sovereignty and jurisdiction over the territorial sea, regulatory control for certain purposes over the contiguous zone, and sovereign rights and related jurisdiction over the natural resources, certain economic activities and environmental protection of the EEZ pursuant to presidential proclamations.

As to those portions of the territorial sea that do not belong to the states, where U.S. jurisdiction over ocean resources is arguably at its strongest, there has been no definitive resolution as to whether U.S. authority over the area amounts to "ownership or control" for purposes of the Antiquities Act. There are, however, several authorities supporting the conclusion that the U.S. does "own" or "control" at least portions of the territorial sea as contemplated by the Antiquities Act. First, the Supreme Court has held that the United States owned the lands within the territorial sea and that the Property Clause of the U.S. Constitution authorized Congress to dispose of such lands.[121] Additionally, prior to President Reagan's proclamation extending U.S. jurisdiction over the seas to parallel international law on the subject, the territorial sea was limited to three miles from the shoreline. During this time, the Supreme Court did state that "[t]here can be no serious question ... that the President in 1949 had the power under the Antiquities Act to reserve the submerged lands and waters within the one-mile belts[122] as a national monument since they were then controlled by the Government of the United States."[123] These holdings support the concept of federal ownership or control and would most likely be applicable to those areas extending twelve miles from the shore which are under federal jurisdiction. The Submerged Lands Act, discussed below, has altered this balance to some extent.

The Submerged Lands Act (SLA)[124] would appear to impact federal ownership and control. The SLA established or confirmed state title to and ownership of "the lands beneath the navigable waters within the boundaries of the respective states and the natural resources within such lands and waters...."[125] The boundaries of the states were designated as generally including the submerged lands up to three geographical miles from each state's coast.[126] The federal government retained, however, "its navigational servitude and rights in and powers of regulation and control of said lands and navigable waters for the constitutional purposes of commerce, navigation, national defense, and international affairs...."[127] Further, these rights are paramount to, but do not include, the states' "proprietary rights of ownership, or the rights of management, administration, leasing, use, and development of the lands and natural resources.... "[128]

Thus, it is clear that, vis-a-vis the states, the federal government does not now "own" the submerged lands and waters to the extent that such rights have been ceded to the states, although the area remains in U.S. ownership vis-a-vis other nations. Determining federal "control" may be somewhat more elusive. Significant federal controls have been both ceded and retained, and, as indicated above, the standard for determining "control" does not appear to have been provided by statute or court decision. However, the Supreme Court's decision in *United States v. California*, which held that the SLA transferred "dominion" over the three mile coastal region to the states and thus nullified a previously valid designation under the Antiquities Act, may support the argument that the President would not have authority to designate a national monument in those areas affected by the SLA.[129] Further, at least one federal court decision has held that the Antiquities Act is inapplicable to state submerged lands, basing its conclusion on *United States v. California*.[130]

Areas within the territorial sea but beyond the region affected by the SLA would appear to remain subject to federal ownership or control as contemplated by the Antiquities Act.[131] Further support for this position is found in the Outer Continental Shelf Lands Act (OCSLA).[132] In describing the OCSLA, the U.S. Supreme Court has stated that "Congress declared that the United States *owned* all submerged land in the continental shelf seaward of the lands granted to the States"[133] However, it would not appear that the Court was interpreting whether the United States had ownership or a lesser interest in Outer Continental Shelf lands in that decision. As this assertion would appear to be nonbinding dicta, the extent of the U.S. claim to submerged lands under the OCSLA remains unclear. Indeed, the OCSLA does not state that the U.S. owns Outer Continental Shelf submerged lands. The statute defines the Outer Continental Shelf as "all submerged lands lying seaward and outside of the area of lands beneath navigable waters as defined in section 1301 of this title, and of which the subsoil and seabed *appertain* to the United States and are subject to its jurisdiction and control."[134] The nature of any U.S. ownership or jurisdictional interest would also appear to be circumscribed by additional provisions of the OCSLA. The law states:

> It is hereby declared to be the policy of the United States that—
> (1) the subsoil and seabed of the outer Continental Shelf appertain to the United States and are subject to its *jurisdiction, control, and power of disposition as provided in this subchapter....*[135]

The use of these terms would appear to indicate that the U.S. interest in Outer Continental Shelf submerged lands does not amount to ownership by virtue of the OCSLA itself, at least as to those areas beyond the territorial sea. However, even beyond the territorial sea, U.S. waters may still be subject to control as contemplated by the Antiquities Act.

As discussed earlier, in its EEZ, a zone beyond its territory, a coastal nation has sovereign rights to explore, exploit, conserve, and manage marine and seabed resources.[136] The Presidential Proclamation establishing the U.S. EEZ states that the United States claims these rights in the 200 nautical mile area allowable under international law.[137] The extent of control claimed by the U.S. over the EEZ is less extensive than that which it has claimed and exercised over the territorial sea. Recognition of a contiguous zone, allows for additional regulatory authority over the area extending 24 nautical miles from the coast for specified purposes. It is not immediately clear if the levels of regulatory authority in these zones would be a determinative factor in deducing if the U.S. exercises "control" under the Antiquities Act. However, it should be noted that the United States claims significant authority, consistent with international law, to regulate the EEZ and contiguous zone for environmental and economic puiposes.[138] Thus, whether control claimed over these submerged lands is sufficient to meet the requirements of the Antiquities Act remains an unsettled issue.

National Park System

While only Congress itself can designate a national park, the National Park Service Organic Act[139] allows the Secretary of the Interior to recommend areas to Congress for inclusion in the National Park System.[140] Upon receiving appropriations for study of specific areas,[141] themselves recommended by DOI, the Secretary must determine and report to Congress whether an area possesses national significance and is suitable and feasible for inclusion in the National Park System.[142] In determining the eligibility of a site for inclusion, the Secretary must consider nine factors:

(i) the rarity and integrity of the resources;
(ii) the threats to those resources;
(iii) [whether] similar resources are already protected in the National Park System or in other public or private ownership;
(iv) the public use potential;
(v) the interpretive and educational potential;

(vi) costs associated with acquisition, development and operation;
(vii) the socioeconomic impacts of any designation;
(viii) the level of local and general public support; and
(ix) whether the area is of appropriate configuration to ensure long-term resource protection and visitor use.[143]

Thus, while the Secretary is directed to evaluate site suitability and while the enumerated factors may militate for or against inclusion of any particular area, there is no indication that marine sites could not be included in the Secretary's recommendations or the National Park System itself. Congress, in fact, has broad power to make such designations and has included marine resources, notably coral reefs, in the National Park System. For example, Congress created the National Park of American Samoa "to preserve and protect the tropical forest and archeological and cultural resources of American Samoa, and of associated reefs."[144] Hence, fishing in the designated areas is prohibited unless for subsistence purposes.[145]

Management of parks is delegated by statute to the National Park Service, which is directed to prepare general management plans. These plans must include, among other things, "measures for the preservation of the area's resources...."[146] Broad general authority to prescribe rules related to the management of the National Park System is also granted to the Secretary of the Interior.[147] Moreover, the Secretary of the Interior is specifically authorized to promulgate rules "concerning boating and other activities on or relating to waters located within areas of the National Park System...."[148]

Thus, while significant legislative action is required for the designation of any national park, marine resources are not outside the scope of the system now in place. Furthermore, the broad management powers generally available under the National Park Service Organic Act and Congress' power to further refine marine resource management could provide flexible and comprehensive regulation.

Because the United States owns and exercises full sovereignty over its territorial seas, establishment of a park in that zone would not appear to pose particular jurisdictional problems, although state ownership, as expressed in the SLA, may require consideration. Designation within the other zones may prove more complicated. The National Park Service Organic Act makes no apparent distinction among waters, and hence ocean resources, based on the various zones. In fact, in 16 U.S.C. § 1 a-2(h), the act states that the Secretary of the Interior shall:

> [p]romulgate and enforce regulations concerning boating and other activities on or relating to waters located within areas of the National Park System, including waters subject to the *jurisdiction* of the United States: *Provided*, That any regulations adopted pursuant to this subsection shall be complementary to, and not in derogation of, the authority of the United States Coast Guard to regulate the use of waters subject to the jurisdiction of the United States.[149]

Thus arguably, under the terms of National Park Service Organic Act, designation of a park in the waters of any zone appears permissible as a form of regulating ocean resources so long as consistent with the sovereignty or jurisdiction that the United States claims.

The National Wildlife Refuge System

The National Wildlife Refuge System Administration Act of 1966 (NWRSAA)[150] authorizes the Secretary of the Interior, acting through the U.S. Fish and Wildlife Service (FWS), "to administer a national network of lands and waters for the conservation, management, and where appropriate, restoration of the fish, wildlife, and plant resources and their habitats within the United States for the benefit of the present and future generations of Americans."[151] Thus, the act would appear to contemplate the possibility of marine resources receiving wildlife refuge status. Nothing in the language of the act would appear to preclude this understanding of the statute. The Refuge System currently includes lakes and marshes and freshwater swamps, certain coastal areas, and submerged lands and waters, although it would appear that each ocean refuge is in some way connected to Hawaiian islands or territories owned by the United States and incorporates the territorial waters of those possessions.[152] Thus, it would not appear that an independent refuge in the EEZ has been established to date.

In some settings, it is clear that the jurisdiction of the FWS over marine resources has been circumscribed. The FWS, as created by the Fish and Wildlife Act of 1956, was composed of the Bureau of Commercial Fisheries and the Bureau of Sport Fisheries and Wildlife.[153] The Bureau of Commercial Fisheries, which had authority over fishery management, was transferred to the Department of Commerce and the National Oceanic and Atmospheric Administration by Executive action.[154] The Bureau of Sport Fisheries and Wildlife remained in DOI, although its authorities related to the protection of migratory marine species of game fish were also transferred by the Reorganization Plan.[155] Under the resulting reorganization, the FWS retained authority over only those Bureau of Commercial Fisheries functions related to "(1) Great Lakes fishery research and activities related to the Great Lakes Fisheries Commission, (2) Missouri River Reservoir research, (3) the Gulf Breeze Biological Laboratory of the said Bureau at Gulf Breeze, Florida and (4) Trans-Alaska pipeline investigations."[156] Thus, it is relatively clear that the FWS does not retain significant jurisdiction over ocean resources under the authorities originally granted to the Bureau of Commercial Fisheries. However, the FWS's jurisdiction under other laws would not appear to be affected by the 1970 reorganization. Indeed, its authority to manage resources as part of the National Wildlife Refuge System was not involved in the transfer of the duties of the Bureau of Commercial Fisheries and, thus, would appear to remain intact to the extent provided for in specific congressional authorizations.

There is no single method for national wildlife refuge designation, and various administrative, executive, and legislative processes have been employed in the past, sometimes in combination.[157] Generally, refuges have been created through legislation, executive order, or acquisition of private land.[158] Past legislation has either directly designated a refuge or authorized specific executive action to do so.[159]

There are two primary statutes guiding FWS regulation of refuge areas, the Refuge Recreation Act of 1962[160] and the National Wildlife Refuge System Administration Act of 1966, significantly amended in 1997. The Secretary is given broad discretion to regulate activities in refuge areas and is authorized to "permit the use of any area within the System for any purpose ... whenever he determines that such uses are compatible with the major purposes for which such areas were established."[161] The act defines compatible use as "a wildlife-dependent recreational use or any other use of a refuge that, in the sound professional judgment of the Director, will not materially interfere with or detract from the fulfillment of the mission of the System or the purposes of the refuge."[162] FWS regulations further clarify the compatible use standard and allow for, under specified circumstances, recreational uses of refuges,[163] certain economic uses,[164] hunting and fishing,[165] and subsistence uses.[166] The regulations also prohibit certain activities.[167] In general, refuge protection standards would appear to be more permissive than the standards applicable to national parks and less permissive than standards applicable to multiple-use lands.[168]

An additional issue is whether the NWRSAA is meant to apply in the territorial sea and the EEZ. The statute states:

> The mission of the System is to administer a national network of lands and *waters* for the conservation, management, and where appropriate, restoration of the fish, wildlife, and plant resources and their habitats *within the United States* for the benefit of present and future generations of Americans.[169]

Units of the Refuge System would presently appear to be confined to areas within the territory of the United States, including the territorial sea.[170] Because of the extent of the U.S. claim to ownership and sovereign jurisdiction over the territorial sea, the territorial sea could arguably be considered "within the United States." On the other hand, it is arguable that "within the United States" does not include any area beyond the U.S. coastline. However, as multiple refuges currently contain portions of the territorial sea and as there is little to indicate that the territorial sea is not meant to be considered a part of the United States for purposes of the NWRSAA, it would seem that refuge designation within the territorial sea is permissible. Whether designation in the EEZ or contiguous zone would be permissible under current law is more questionable.

Conforming to international law, the Presidential Proclamation establishing the EEZ, states that the EEZ "remains an area *beyond the territory* and territorial sea of the United States in which all States enjoy the high seas freedoms of navigation, overflight, the laying of submarine cables and pipelines, and other internationally lawful uses of the sea."[171] As pointed out in *Natural Resources Defense Council v. United States Department of the Navy*, "while the EEZ is not part of United States territory, the United States does enjoy certain 'sovereign rights' there, including sovereign rights 'for the purpose of exploring, exploiting, conserving and managing natural resources."[172] Thus, while the United States would likely be empowered to create refuge-like protections for areas within its EEZ, if Congress deemed it appropriate, it is doubtful that the areas beyond the territorial sea would qualify as "within the United States" for current NWRSAA purposes.

Marine Preservation through Executive Order

Near the end of his administration, President Clinton issued Executive Orders 13,178[173] and 13,196,[174] thereby creating the Northwestern Hawaiian Islands Coral Reef Ecosystem Reserve (NHIR). The reserve covers approximately 99,500 square nautical miles, making it substantially larger than other marine sanctuaries. NHIR would appear to be the first U.S. designation of its kind, i.e. the creation of a "marine reserve" via executive order. Like other MPAs discussed in this report, the reserve surrounds certain U.S.-owned Pacific islands.

Presidential designation of the NHIR was specifically authorized in the 2000 amendments to the National Marine Sanctuaries Act,[175] which were enacted shortly before President Clinton issued the NHIR Orders. The amended Act also directs the Secretary of Commerce, upon the creation of a reserve, to initiate its designation as a marine sanctuary under the National Marine Sanctuaries Act.[176] Further, the amendments state that, in the period before designation of the reserve as a marine sanctuary, the NHIR be managed in a manner consistent with the purposes of the National Marine Sanctuaries Act.[177] These requirements are reflected in the Executive Orders as well.[178] Thus far, the NHIR has been designated and the process for redesignation as a marine sanctuary has begun as well. It should be noted, however, that unless the provisions of an executive order are enacted into law, an executive order can be delayed, abandoned, or expressly overturned by a later order. The congressional authorization here does not compel a designation, and while it authorizes the creation of a reserve, it would not appear to imbue the Clinton Orders with any additional force of law.

The regulatory authority exercised over the NHIR pursuant to the Executive Orders derives from several preexisting legal sources. Generally, an executive order is used by the President to direct some action within the executive branch. Executive orders may be based upon the President's constitutional powers and/or upon specific statutory authority. The Orders at issue here cite several authorities in support of the President's designation, including the Constitution, the NMSA, Magnuson-Stevens, the Marine Protection, Research, and Sanctuaries Act, the CZMA, the Endangered Species Act, the Marine Mammal Protection Act, the Clean Water Act, the National Historic Preservation Act, the NWRSAA, and "other pertinent statutes."[179] Thus, President Clinton directed the federal agencies to protect the NHIR as authorized by the applicable statutes. The Executive Orders would not appear to authorize or establish any new regulatory capabilities that were not available under preexisting law.

The Orders themselves establish certain baseline protections for the designated area. They limit development of the reserve's resources and generally prohibit oil and gas exploration and production, anchoring on coral reefs, alteration of the seabed, discharges into the reserve, and, under certain circumstances, the taking of biological resources.[180] The Orders permit certain categories of commercial and recreational fishing to continue at current levels in most portions of the reserve, while establishing eight Preservation Areas where permitted fishing may temporarily take place, and seven Preservation Areas where no resource development is allowed.[181]

The Orders also provide for additional, more detailed management and conservation measures to be developed by the Secretary of Commerce in conjunction with the Secretary of the Interior, the State of Hawaii, the Western Pacific Fishery Management Council (WesPac),[182] and the Coral Reef Ecosystem Reserve Council. This group is directed to develop general Reserve Management Principles, conservation measures, and a Reserve Operations Plan. The basic regulatory authority contained in several of the most relevant statutory authorities has been described above; however, several additional regulatory authorities deserve attention here: namely the Endangered Species Act, the Marine Mammal Protection Act, the Clean Water Act, and the Ocean Dumping Act.

The Endangered Species Act (ESA)[183] prohibits the "taking"[184] of any threatened or endangered species and generally prohibits federal agencies from harming these species though direct action or through federally funded activities.[185] Species protection under the

ESA has been interpreted to include the designation and protection of critical habitat.[186] Thus, significant protections for both individual species and their habitat may be available under the ESA. However, the ESA allows for "take permits," which could include allowances for harm to individuals or critical habitat if it is determined that the taking is incidental to a lawful activity and will not endanger the species.[187]

The Marine Mammal Protection Act (MMPA)[188] is also cited as an authority for the President's reserve designation. Unlike the ESA and its protections for critical habitat designation, the MMPA addresses only species management. The law generally prohibits the taking of marine mammals but allows for incidental takings during fishing operations. It does authorize the Secretary of Commerce to close fisheries or revoke individual permits if the terms of incidental take restrictions are not followed.[189]

The Clean Water Act (CWA)[190] prohibits discharging pollutants into ocean waters when such discharges might cause adverse impacts to the marine environment.[191] The act directs the Administrator of the EPA to "promulgate guidelines for determining the degradation of the waters of the territorial seas, the contiguous zone, and the oceans," incorporating "the effect of disposal of pollutants on marine life."[192] Using these guidelines, the appropriate Regional, State, or Tribal Director[193] must determine if any pollutant discharge will unreasonably degradel[194] the marine environment and can issue a discharge permit only upon a finding that unreasonable degradation will not occur.[195]

Title I of the Marine Protection, Research and Sanctuaries Act (MPRSA),[196] commonly referred to as the Ocean Dumping Act, supplements the CWA's discharge limitations and prohibits the "dumping in ocean waters of any material which would adversely affect human health, welfare, or amenities, or the marine environment, ecological systems, or economic potentialities," without appropriate authorization.[197] The MPRSA authorizes EPA to designate specific areas where dumping is completely prohibited and to issue discharge permits, similar to those under the CWA.[198] Permit issuance is governed by the CWA's ocean discharge criteria and the London Dumping Convention.[199]

Additionally, while the Executive Orders direct coordinated management and interagency cooperation, the overlapping agency jurisdiction could result in some level of tension. While in most instances the Executive Orders appear to vest primary management authority in the Secretary of Commerce pursuant to congressional authorization, the Secretary would not appear to have primary jurisdiction over portions of the NHIR. The reserve boundaries are established in Executive Order 13,178, stating:

> The Reserve shall be adjacent to and seaward of the seaward boundaries of the State of Hawaii and the Midway Atoll National Wildlife Refuge, and *shall overlay the Hawaiian Islands National Wildlife Refuge* to the extent that it extends beyond the seaward boundary of the State of Hawaii.[200]

Thus, a portion of the NHIR includes a portion of a National Wildlife Refuge. The Secretary of Commerce's management authority may be problematic in this area because of a general statutory requirement that the National Wildlife Refuge System be managed by the Secretary of the Interior through the FWS.[201] However, as indicated in the legislative history surrounding the development of this jurisdictional limitation and the cases interpreting it, DOI may delegate some of its responsibilities and coordinate management activities with state or federal agencies.[202] The courts have indicated that the level of discretion that can be vested in other entities with regard to refuge management is limited and that DOI must retain ultimate authority and responsibility for refuge resources.[203]

It is arguable that the law authorizing presidential designation of the NHIR had some impact on the general placement of refuge authority in DOI. The law states that the reserve is "to be managed by the Secretary of Commerce."[204] This could be interpreted as an authorization for the President to effectively place primary management authority for a Wildlife Refuge in the hands of the Secretary of Commerce should the President choose to include such an area in the reserve. As described above, this could result in the application of the generally weaker reserve protective standards to the refuge. On the other hand, it is arguable that had Congress intended such a result, it would have spoken to this issue directly. There is a general presumption that Congress will specify its intention that a new statute is meant to supersede an earlier one.[205] Further, courts will generally read a subsequent enactment as an amendment or repeal only when the conflict between two provisions is irreconcilable or the subsequent enactment is clearly intended as a substitute.[206] Applying this general canon of statutory construction would appear to favor an interpretation of the reserve designation provision that leaves the primary refuge management authorities with FWS alone. Still, the statute arguably authorizes cooperative management efforts among the relevant agencies, even if primary authority remains with DOI.

The Executive Orders would not appear to definitively clarify where management authority rests. They direct that "the Secretary of Commerce, or his designee, ... manage the Reserve," and that the management system facilitate coordination among the state and federal agencies involved.[207] These statements appear to indicate an intention that the Secretary of Commerce have ultimate regulatory responsibility. On the other hand, there is some indication that regulatory authority over the Hawaiian Islands National Wildlife Refuge is separate from the rest of the reserve. For example, the provisions that guide the preparation of the Reserve Operations Plan (ROP) state that the ROP must provide for coordinated management between the Reserve and the Hawaiian Islands National Wildlife Refuge.[208] Similarly, the Orders direct the Secretary of Commerce to negotiate any necessary Memoranda of Understanding with the Secretary of the Interior and the state of Hawaii regarding management coordination between the reserve and the Wildlife Refuge.[209] Thus, arguments could be made that ultimate Refuge authority has been placed in either of the agencies by the Executive Orders. However, as described above, should a court find that the congressional authorization of the creation of a reserve was not intended to supersede the general requirement that refuge authority rest in DOI, an Executive Order would not appear to be the appropriate vehicle for redistributing management responsibilities, except to the extent afforded by established case law.

The enabling statute and the Executive Orders also leave the role of the regional fishery management council, WesPac, vis-a-vis the other regulatory authorities relatively vague. It would appear, however, that WesPac is intended to remain an active regulator under Magnuson-Stevens and continue with such activities as it has heretofore undertaken. As described above, WesPac, as the regional fishery management council, is primarily responsible for the development of fishery management plans in accordance with Magnuson-Stevens, subject to approval by the Secretary of Commerce.[210] WesPac has apparently voiced some concern that certain requirements contained in the Executive Orders and the NHIR regulations may violate the Magnuson Act by preventing resource utilization as permitted under law.[211] This opinion may result from an understanding that Magnuson-Stevens generally requires fishery management plans to include conservation and management measures that achieve "optimum yield" from each fishery while preventing overfishing.[212] These management measures are to be based on the best scientific information available.[213] Thus, it is conceivable that the base restrictions required by the Executive Orders could conflict with the levels of sustainable fishing as determined by the regional council. However, the maximum sustainable yield from any fishery is to be reduced by the relevant social, economic, and ecological factors.[214] Further, conservation and management activities are required to rebuild, restore, or maintain fishery resources and the marine environment.[215] It would appear to be within the President's authority to direct fishery management so that conservation of the marine environment is the primary objective. If, however, specific restrictions on resource exploitation could not be supported as necessary by scientific information, as required by law, additional fishing may have to be permitted, unless the more stringent restrictions could be based on some superseding authority.

Conclusion

In light of the recent publication of the Revised Draft Framework for Developing the National System of Marine Protected Areas and the introduction of H.R. 21 in the 110th Congress, it is important to understand the statutory and regulatory background for the designation of MPAs. A series of statutory authorities exist for the creation and management of MPAs. The National Marine Sanctuaries Act would appear to be directly aimed at the creation of MPAs, although, in practice, the protections provided areas designated under its authority have not necessarily been extensive. Other legislation, such as the Coastal Zone Management Act and the Magnuson-Stevens Fishery Conservation and Management Act, also have the potential to protect marine resources in a fashion similar to MPAs. More general preservation laws may also be an option for the protection and management of marine resources. The Antiquities Act, the National Park Service Organic Act, and the National Wildlife Refuge System Administration Act appear to generally allow the designation of marine resources as national monuments, national parks, or national wildlife refuges. However, use of these various conservation authorities for the creation of an MPA would carry concerns peculiar to each particular statute and would be limited by the extent of U.S. jurisdiction over offshore lands and waters. Their application to the territorial sea would generally appear permissible. Application beyond the territorial sea is less certain.

End Notes

[1] The Committee is to

(a) coordinate the activities of executive departments and agencies regarding ocean-related matters in an integrated and effective manner to advance the environmental, economic, and security interests ofpresent and future generations of Americans; and

(b) facilitate, as appropriate, coordination and consultation regarding ocean- related matters among Federal, State, tribal, local governments, the private sector, foreign governments, and international organizations. Exec. Order No. 13366, 69 Fed. Reg. 76,591 (December 17, 2004).

[2] U.S. Ocean Action Plan: The Bush Administration's Response to the U.S. Commission on Ocean Policy, (December 17, 2004) available at [http://ocean.ceq.gov/].

[3] H.R. 21, a bill introduced in the 110th Congress that is intended to "provide a national policy for our oceans," is one manifestation of this debate.

[4] Act of October 23, 1972, P.L. 92-532 (codified at scattered sections of titles 16 and 33). 516 U.S.C. §§ 1801-1882.

[5] 16 U.S.C. §§ 1801-1882.

[6] 16 U.S.C. §§ 1451-1465.

[7] Exec. Order No. 13,158, 65 Fed. Reg. 34,909 (May 26, 2000).

[8] *Id.*

[9] *Id.* at § 3 (emphasis added).

[10] *Id.* at § 7.

[11] The Revised Draft Framework is available at [http://www.mpa.gov/pdf/national-system/ revise draft frmwk 0308.pdf]. The comment period ends May 16, 2008.

[12] Draft Framework, at 1-2.

[13] Draft Framework, at 5.

[14] *Id.*

[15] Relevant measurements: (geographical mile = 6,087.15 ft.), (land mile = 5,280 ft.),(marine league = 18,228.3 ft.), (nautical mile = 6,076.1 ft.).

[16] United Nations Convention on the Law of the Sea, December 10, 1982, arts. 2.1, 2.2, 3, 21 I.L.M. 1261 (entered into force November 16, 1994).

[17] *Id.* at art. 33.

[18] *Id.* at art. 56.1.

[19] *Id.*; Proclamation 5030, Exclusive Economic Zone of the United States of America, March 10, 1983.

[20] *See* UNITED NATIONS, STATUS OF THE UNITED NATIONS CONVENTION ON THE LAW OF THE SEA 10 (May 30, 2002); Exclusive Economic Zone of the United States of America, Proclamation No. 5030, 48 Fed. Reg. 10,605 (March 14, 1983); Territorial Sea of the United States of America, Proclamation No. 5928, 54 Fed. Reg. 777 (December 27, 1988); Contiguous Zone of the United States, Proclamation No. 7219, 64 Fed. Reg. 48,701 (August 2, 1999).

[21] See, e.g., 16 U.S.C. §§ 1362(15); 1432(3); 1538(a)(1)(B)-(C).

[22] U.S. CONST. art IV, § 3.

[23] U.S. CONST. art. I, § 8.

[24] U.S. CONST. art II, § 2.

[25] See CRS Report RL32185, U.*N. Convention on the Law of the Sea: Living Resources Provisions*, by Eugene Buck.

[26] Protocol Amending the 1916 Convention for the Protection of Migratory Birds (December 5, 1995), available at [http://www.le.fws.gov/pdffiles/Canada Mig Bird Treaty.pdf].

[27] UNCLOS III arts. 2.1, 17-26, 37-44, 53.

[28] Proc. No. 5928 (December 27, 1988).

[29] See, e.g., United States v. California, 436 U.S. 32, 36 (1978); United States v. Alaska, 422 U.S. 184, 199 (1975).

[30] See UNCLOS III, art. 2.1; Restatement (Third) of the Foreign Relations Law of the United States, §§ 512, 513 (1986).

[31] 43 U.S.C. §§ 1301-1303, 1311-1315.

[32] Relevant measurements: (geographical mile = 6,087.15 ft.), (land mile = 5,280 ft.), (marine league = 18,228.3 ft.), (nautical mile = 6,076.1 ft.).

[33] 43 U.S.C. § 1301(a)(2). Certain coastal states, namely Florida and Texas, have slightly different claims, as recognized by the Submerged Lands Act. Each of these states may claim title to 3-marine league (9 nautical miles) seaward boundary within which the states own and regulate marine resources, including fish and offshore minerals. *Id.* See also United States v. Louisiana, 363 U.S. 1, 66 (1960) ("pursuant to the Annexation Resolution of 1845, Texas' maritime boundary was established at three leagues from its coast for domestic purposes .... Accordingly, Texas is entitled to a grant of three leagues from her coast under the Submerged Lands Act"); United States v. Florida, 363 U.S. 121, 129 (1960) ("We hold that the Submerged Lands Act grants Florida a three-marine-league belt of land under the Gulf, seaward from its coastline, as described in Florida's 1868 Constitution.").

[34] 43 U.S.C. § 1314(a).

[35] 43 U.S.C. § 1311(a)(2).

[36] 43 U.S.C. § 1302.

[37] Proc. No. 5030 (March 10, 1983).

[38] UNCLOS III Arts. 56, 58.

[39] *Id.* at art. 56.1.

[40] *Id.* at art. 56.1(b).

[41] *Id.* at § 514, comment. c.

[42] *Id.* at art. 33.

[43] Proc. No. 7219 (September 2, 1999).

[44] *Id.*

[45] United States v. De Leon, 270 F.3d 90, 91 n.1 (Pt Cir. 2001).

[46] *See* Vermilya-Brown Co. v. Connell, 335 U.S. 377, 381 (1948); Cuban Am. Bar Ass'n v. Christopher, 43 F.3d 1412, 1425 (11th Cir.1995) (control and jurisdiction is not equivalent to sovereignty).

[47] 16 U.S.C. §§ 1431-1445b (2003)(as amended).

[48] 16 U.S.C. § 1431(b)(2).

[49] *Id.* 16 U.S.C. §§ 1431(a), 1433.

[50] COMMITTEE ON THE EVALUATION, DESIGN, AND MONITORING OF MARINE RESERVES AND PROTECTED AREAS IN THE U.S., NAT 'L RESEARCH COUNCIL, MARINE PROTECTED AREAS: TOOLS FOR SUSTAINING OCEAN ECOSYSTEMS at 156 (2001).

[51] See [http://sanctuaries.noaa.gov/visit/welcome.html].

[52] Congress has designated two marine sanctuaries. See Florida Keys National Marine Sanctuary and Protection Act, P.L. 101- 605, (1990); Hawaiian Islands National Marine Sanctuary Act, P.L. 102-587, (1992). The others were created administratively.

[53] 16 U.S.C. § 1433(a).

[54] 16 U.S.C. § 1434(a)(1)(A).

[55] 16 U.S.C. § 1434(a)(2)(C)(i). "The terms of designation of a sanctuary shall include the geographic area proposed to be included within the sanctuary, the characteristics of the area that give it conservation, recreational, ecological, historical, research, educational, or esthtic [sic] value, and the types of activities that will be subject to regulation by the Secretary to protect those characteristics. The terms of designation may be modified only by the same procedures by which the original designation is made." 16 U.S.C. § 1434(a)(4).

[56] 16 U.S.C. § 1434(a).

[57] 16 U.S.C. § 1434(a)(3).

[58] 16 U.S.C. § 1434(a)(6).

[59]16 U.S.C. § 1433(a).

[60] 16 U.S.C. §§ 1433(b), (b)(2).

[61] 16 U.S.C. § 1434(b)(1).

[62] *Id.*

[63] 16 U. S. C . § 1431(b)(6).

[64] Robin Kundis Craig, Taking Steps Toward Marine Wilderness Protection? Fishing and Coral Reef Marine Reserves in Florida and Hawaii, 34 MCGEORGE L. REV. 155, 204 (2003).

[65] 16 U.S.C. § 1433(b)(3). The National Environmental Policy Act (NEPA), 42 U.S.C. §§ 4321 *et seq.*, requires the preparation of a detailed statement as to the environmental effects of any major federal actions that might significantly affect the environment.

[66] 16 U.S.C. §§ 1433(b)(1)(H), (I).

[67] 16 U.S.C. § 1433(b)(1)(C).

[68] See discussion in the next section of this Report.

[69] 16 U.S.C. § 1434(a)(5).

[70] *Id.*

[71] 16 U.S.C. § 1434(c).

[72] 16 U.S.C. § 1441(g).

[73] 16 U.S.C. § 1443(b).

[74] 16 U.S.C. § 1443(d).

[75] United States v. MN Jacquelyn L, 100 F.3d 1520 (11th Cir. 1996); see also 16 U.S.C. §§ 1443(a)(1), (2).

[76] 16 U.S.C. §§ 1801-1882.

[77] 16 U.S.C. § 1801(a)(6).

[78] 16 U.S.C. §§ 1801(a)(7), 1811; *see supra*, note 30 and accompanying text.

[79] 16 U.S.C. § 1852(a).

[80] 16 U.S.C. §§ 1853, 1852(h), 1851(a)(1)-(7).

[81] 16 U.S.C. § 1851(a)(1).

[82] 16 U.S.C. § 1802(28), as amended by P.L. 104-297, § 102.

[83] 16 U.S.C. § 1802(29).

[84] 16 U.S.C. § 1854. Approval authority has been delegated from the Secretary to the NOAA Administrator. The Administrator subsequently delegated approval authority to the National Marine Fisheries Service (NMFS) Director, who in turn delegated to the regional NMFS directors. Each delegation required the subordinate agency or council to "advise" the delegator prior to approving an FMP or FMP amendment. The Court of Appeals for the D.C. Circuit held in 1991 that each agency retained approval authority under these delegations. C&W Fish Company, Inc. v. Fox, 931 F.2d 1556 (D.C. Cir. 1991). Since the time of that decision, the directives establishing the delegations in question have been modified, removing certain language relied upon by the court in reaching its conclusion. However, the term "advise" remains in place and was also integral to the court's decision that superior agency approval and not mere notice was required. Subsequent decisions do not appear to have addressed this issue. Thus it remains, for the time being, unclear whose approval of FMPs or FMP amendments is necessary.

[85] 16 U.S.C. §§ 1851(a)(6), 1853(b)(12). The options suggested by Magnuson-Stevens include permitting, specified zones, overall catch limits, and vessel and gear prohibitions. 16 U.S.C. § 1853.

[86] Fikret Berkes, Robin Mahon, Patrick McConney, Richard Pollnac, and Robert Pomeroy, MANAGING SMALL-SCALE FISHERIES: ALTERNATIVE DIRECTIONS AND METHODS 6.3.2 (2001).

[87] 16 U.S.C. § 1853(a). This section states that FMPs must "specify objective and measurable criteria for identifying when the fishery to which the plan applies is overfished ... and, in the case of a fishery which the Council or the Secretary has determined is approaching an overfished condition or is overfished, contain conservation and management measures to prevent overfishing or end overfishing and rebuild the fishery...." *Id.* See also 50 C.F.R. § 600.1002.

[88] See, e.g., 50 C.F.R. § 300.29 (eastern Pacific fisheries management); 50 C.F.R. § 622.34 (Gulf EEZ seasonal and/or area closures); 50 C.F.R. § 600.51 (foreign fishing closure procedures).

[89] 16 U.S.C. § 1855(b); 50 C.F.R. § 600.805(a).

[90] 16 U.S.C. §§ 1855(b), 1882(a); 50 C.F.R. § 600.805(a).

[91] 50 C.F.R. § 600.815.

[92] 50 C.F.R.§§ 600.815(a)(2)(ii)(E), 600.810(a).

[93] 50 C.F.R. § 600.815(a)(2)(ii).

[94] 50 C.F.R. § 600.815.

[95] 16 U.S.C. § 1852(h).

[96] 16 U.S.C. § 1853(c)(2); 50 C.F.R. § 600.315(d).

[97] 50 C.F.R. § 600.335(c)(2), (d).

[98] 43 U.S.C. §§ 1331-1356.

[99] 43 U.S.C. §§ 1301, 1331(a).

[100] 43 U.S.C. §§ 1334(a), 1337.

[101] 43 U.S.C. § 1344.

[102] *Id.*

[103] 43 U.S.C. § 1334; The OCSLA states:

> The Secretary may at any time prescribe and amend such rules and regulations as he determines to be necessary and proper in order to provide for the prevention of waste and conservation of the natural resources of the outer Continental Shelf, and the protection of correlative rights therein, and, notwithstanding any other provisions herein, such rules and regulations shall, as of their effective date, apply to all operations conducted under a lease issued or maintained under the provisions of this subchapter.

[104] 43 U.S.C. § 1340(c)(1).

[105] 43 U.S.C. §§ 1340(c), 1334(a)(2) (stating that permanent cancellation of a lease or permit may occur provided that certain findings are made and appropriate administrative procedures are followed); see also Mobil Oil Exploration and Producing Southeast, Inc. v. U.S., 530 U.S. 604 (2000); Gulf Oil Corp. v. Morton, 493 F.2d 141 (9th Cir. 1973).

[106] 43 U.S.C. § 1340(c)(2).

[107] *Id.*

[108] It should be noted that if a lease has been granted by the government and subsequent development is not permitted, the denial of the permit may constitute a compensable taking. See Mobil Oil Exploration and Producing Southeast, Inc. v. U.S., 530 U.S. 604 (2000).

[109] *See supra*, note 31 and accompanying text.

[110] 16 U.S.C. §§ 1451-1465.

[111] 16 U.S.C. § 1452(1), (2).

[112] 16 U.S.C. § 1455(d)(2), (9)-(12).

[113] 16 U.S.C. §§ 1455(d)(16).

[114] 16 U.S.C. § 1456(c).

[115] Act of June 8, 1906, codified at 16 U.S.C. §§ 431-433.

[116] 16 U.S.C. § 431 (2003) (emphasis added).

[117] See generally Sanjay Ranchod, The *Clinton National Monuments: Protecting Ecosystems with the Antiquities Act*, 25 HARV. ENVTL. L. REV. 535, 540 (2001).

[118] See Proclamation No. 2281, 52 Stat. 1541 (April 26, 1938) (Channel Islands National Park); Proclamation No. 2337, 3 C.F.R. 88 (1938-1943) (Santa Rosa Island National Monument); Proclamation No. 3443, 3 C.F.R. 152 (1959-1963) (Buck Island Reef National Monument); Proclamation No. 7392, 66 Fed. Reg. 7,335, 7,336 (January 17, 2001) (Buck Island Reef National Monument Expansion); Proclamation No. 7399, 66 Fed. Reg. 7,364 (January 17, 2001) (Virgin Islands Coral Reef National Monument).

[119] Ranchod, *supra note* 116, at 578-79; see also Cappaert v. United States, 426 U.S. 128, 138-42 (1976); United States v. Oregon, 295 U.S. 1, 14 (1935); United States v. California, 436 U.S. 32, 36 n.9 (1978). See Cameron v. United States, 252 U.S. 450 (1920); Wyoming v. Franke, 58 F. Supp. 890 (D. Wyo. 1945); Alaska v. Carter, 462 F. Supp. 1155 (D. Alaska 1978); Anaconda Copper Co. v. Andrus, 14 Env't Rep. Cas. (BNA) 1853 (D. Alaska 1980).

[120] Proclamation No. 7392, 66 Fed. Reg. 7,335, 7,336 (January 17, 2001).

[121] United States v. California, 332 U.S. 19 (1947); Alabama v. Texas, 347 U.S. 272, 273-74 (1954).

[122] The decision speaks in terms of one-mile belts because this was the location of the national monument and submerged lands at issue in the case. It would not appear, however, that the reasoning behind the decision would be limited to a one-mile region within the territorial sea. United States v. California, 436 U.S. 32, 34-35 (1978).

[123] United States v. Calilfornia, 436 at 36.

[124] Act of May 22, 1953, ch. 65, 67 Stat. 29 (codified at 43 U.S.C. §§ 1301-1303, 1311¬1315).

[125] 43 U.S.C. § 1311.

[126] 43 U.S.C. § 1312; *see supra*, note 30 and accompanying text.

[127] 43 U.S.C. § 1314.

[128] *Id.*

[129] United States v. California, 436 U.S. 32, 36 (1978).

[130] Subaqueous Exploration & Archaeology Ltd. v. Unidentified, Wrecked and Abandoned Vessel, 577 F. Supp. 597, 610 (D. MD 1983).

[131] See United States v. California, 436 U.S. 32, 36 (1978). At the time of the designation, which the court held was then within the President's authority, the territorial sea only extended three miles from shore. The subsequent expansion of the territorial sea would not appear to fundamentally alter the type of control asserted over the ocean resources, just the geographic scope of control. Thus, it seems reasonable to conclude that the court's reasoning in determining that the Antiquities Act applied within the three-mile area would now apply to areas outside the three mile region but still within the twelve mile boundary.

[132] 43 U.S.C. §§ 1331 *et seq.*

[133] U.S. v. California, 381 U.S. 139, 148 (1965) (emphasis added).

[134] 43 U.S.C. § 1331(a) (emphasis added).

[135] 43 U.S.C. § 1332 (emphasis added).

[136] Proclamation No. 5030 (March 10, 1983), *reprinted in* 16 U.S.C. § 1453 note.

[137] *Id.*

[138] See UNCLOS III arts. 61-73. Cases deciding the Antiquities Act did not apply beyond the three mile point were decided before the extension of the territorial sea and claim to the contiguous zone and EEZ. See Treasure Salvors, Inc. v. Unidentified Wrecked and Abandoned Sailing Vessel, 569 F.2d 330 (5th Cir. 1978).

[139] 16 U.S.C. §§ 1-3.

[140] 16 U.S.C. § la-5.

[141] The Secretary may initiate the listing process without specific congressional authorization when such activities cost less than $25,000. *Id.* §§ la-5(b)(3), (4).

[142] 16 U.S.C. § la-5(c).

[143] 16 U.S.C. § 1 a-5(c)(3)(A).

[144] 16 U.S.C. § 410qq(b); *see also* 16 U.S.C. §§ 398 et seq. (Virgin Islands National Park).

[145] 16 U.S.C. § 410qq-2(b)(2).

[146] 16 U.S.C. § la-7(b).

[147] 16 U.S.C. § 3.

[148] 16 U.S.C. § la-2(h) (emphasis in original).

[149] 16 U.S.C. § la-2(h) (emphasis added).

[150] 16 U.S.C. §§ 668dd-668ee.

[151] 16 U.S.C. §§ 668dd(a)(1), (2).

[152] See [http://www.fws.gov/data/NWRdata.htm].

[153] Reorg. Plan No. 3 of 1940.

[154] Reorg. Plan No. 4 of 1970, 35 *Fed. Reg.* 15,627 (October 3, 1970). The jurisdiction of the FWS as to marine resources has changed dramatically over the years. In 1939, the Bureau of Fisheries within the Department of Commerce was transferred to DOI, and later incorporated into the FWS. Subsequently, the FWS Bureau of Commercial Fisheries was returned to the Department of Commerce, and renamed the National Marine Fisheries Service. *See* George Cameron Coggins and Doris K. Nagel, Nothing Beside Remains: The Legal Legacy of James G. Watt's Tenture [sic] as Secretary of the Interior on Federal Land Law and Policy, 17 B.C. ENVTL. AFF. L. REV. 473, 550 n. 46 (Spring 1990).

[155] Reorg. Plan No. 4 of 1970, 35 Fed. Reg. 15627 (Oct 3, 1970).

[156] *Id.*

[157] Richard J. Fink, The National Wildlife Refuges: Theory, Practice, and Prospect, 18 HARV. ENVTL. L. REV. 1, (1994).

[158] The primary mechanism for expanding the National Wildlife Refuge System is acquisition of private lands. The most often utilized law of this sort is the Migratory Bird Conservation Act (MBCA) of 1929. In addition, the Fish and Wildlife Coordination Act, the Fish and Wildlife Act of 1956, and the Endangered Species Act also provide general acquisition authorities; these statutes rely on annual appropriations from Congress.

[159] See, e.g., Act of May 23, 1908, P.L. 136, ch. 192 (codified at 16 U.S.C. § 671).

[160] 16 U.S.C. §§ 460(k)-460(k)(4).

[161] 16 U.S.C. § 668dd(d)(1).
[162] 16 U.S.C. § 668ee(1).
[163] 50 C.F.R. § 26 (2003) (including special regulations for individual refuges and generally applicable provisions which cover the process for determining which uses are compatible uses).
[164] 50 C.F.R. § 29.
[165] 50 C.F.R. § 32.
[166] See, e g 50 C.F.R. §§ 36.11-36.16; 100.
[167] 50 C.F.R. §§ 27, 70.
[168] Richard J. Fink, The National Wildlife Refuges: Theory, Practice, and Prospect, 18 HARV. ENVTL. L. REV. 1, 27 (1994).
[169] 16 U.S.C. § 668dd(a)(2) (emphasis added).
[170] See [http://www.fws.gov/data/NWRdata.htm].
[171] Pres. Proc. No. 5030 (March 10, 1983) (emphasis added).
[172] Natural Resources Defense Council v. United States Department of the Navy, No. CV- 01-07781 CAS(RZX) (C.D. Cal., September 17, 2002) (internal citations omitted).
[173] Exec. Order No. 13,178, 65 Fed. Reg. 76,903 (December 7, 2000).
[174] Exec. Order 13,196, 66 Fed. Reg. 7,395 (January 23, 2001) (finalizing and slightly modifying EO 13,178 after public comment.)
[175] Act of November 13, 2000 P.L. 106-513; see also Act of December 21, 2000, P.L. 106-554, § 1(a)(4).
[176] *Id.*
[177] *Id.* at § 6(g), *codified at* 16 U.S.C. § 6401 Note.
[178] See, e.g., *Exec.* Order No. 13,178; Exec. Order 13,196.
[179] Exec. Order 13,178 preamble.
[180] Exec. Order No. 13,178 § 7(b)(1)-(5); Exec. Order 13,196, § 3(4).
[181] Exec. Order 13,196 §§ 3-4, amending Exec. Order 13,178 §§ 7-8.
[182]This is the applicable regional fishery council under Magnuson-Stevens.
[183] 16 U.S.C. §§ 1531-1544 (1994).
[184] "Take" means "to harass, harm pursue, hunt, shoot, wound, kill, trap, capture, or collect, or attempt to engage in any such conduct," and includes significant habitat modification or degradation that actually kills or injures wildlife. 16 U.S.C. § 1453(19).
[185] 16 U.S.C. §§ 1538(a), 1536(a).
[186] See Babbitt v. Sweet Home Chapter of Communities for a Great Oregon, 515 U.S. 687, 708 (1995).
[187] 16 U.S.C. § 1536(b)(4).
[188] Marine Mammal Protection Act, 16 U.S.C. §§ 1361- 1421.
[189] 16 U.S.C. § 1387.
[190] 33 U.S.C. §§ 1251 *et seq.*
[191] 33 U.S.C. § 1343(a), (c).
[192] *Id.*
[193] According to EPA regulations, the term "Director" may have several different meanings depending upon the situation. The regulations state:
Director means the Regional Administrator or the State Director, as the context requires, or an authorized representative. When there is no "approved State program," and there is an EPA administrative program, "Director" means the Regional Administrator. When there is an approved State program, "Director" normally means the State Director. In some circumstances, however, EPA retains the authority to take certain actions even when there is an approved State program. (For example, when EPA has issued an NPDES permit prior to the approval of a State program, EPA may retain jurisdiction over that permit after program approval, see § 123.1.) In such cases, the term "Director" means the Regional Administrator and not the State Director. 40 C.F.R. § 122.2.
[194] Unreasonable degradation is defined as "(1) Significant adverse changes in ecosystem diversity, productivity and stability of the biological community within the area of discharge and surrounding biological communities, (2) Threat to human health through direct exposure to pollutants or through consumption of exposed aquatic organisms, or (3) Loss of esthetic, recreational, scientific or economic values which is unreasonable in relation to the benefit derived from the discharge."40 C.F.R. § 125.121(e) (2003).
[195] 40 C.F.R. §§ 125.122, 125.123 (2003).
[196] Act of October 23, 1972, P.L. 92-532 (codified at 33 U.S.C. §§ 1401-1445).
[197] 33 U.S.C. §§ 1411(a)(1)-(2), 1402.
[198] 33 U.S.C. § 1412(c).
[199] 33 U.S.C. §§ 1343(c), 1412(a).
[200] Exec. Order 1378 at § 3. It should also be noted that the precise boundaries of the Hawaiian Islands National Wildlife Refuge are relatively unclear. The Refuge was established by President Theodore Roosevelt in 1909 by Executive Order and subsequently included in the National Wildlife Refuge System. The boundaries are not precisely described in the original Executive Order and would appear to be based solely upon a crudely

drawn map. The FWS has apparently established working boundaries, based on practical enforcement of its regulations; however, should jurisdictional authority among the agencies become an issue, precise delineation of refuge boundaries may become necessary.

201 16 U.S.C. § 668dd(a)(1); see also S. Rep. No. 94-593, at 2 (1976)(indicating that split jurisdiction should be avoided when possible).

202 H.Rept. 104-218, at 12 (1995).

203 Calhoun County v. United States, 132 F.3d 1100, 1102 (5th Cir. 1998); Bunch v. Hodel, 793 F.2d 129 (6th Cir. 1986); Trustees for Alaska v. Watt, 524 F. Supp. 1303, 1304-05 (D. Alaska 1981).

204 Act of December 21, 2000, P.L. 106-554, § 1(a)(4) (Div. B, Title I, § 144(f)); Act of November 13, 2000, P.L. 106-513, § 6(g).

205 United States v. Fausto, 484 U.S. 439, 453 (1988).

206 Watt v. Alaska, 451 U.S. 259, 266 (1981); Posadas v. National City Bank, 296 U.S. 497, 503 (1936).

207 Exec. Order 13,178 at § 4.

208 *Id.* at § 5(b)(1).

209 *Id.* at § 5(d).

210 16 U.S.C. § 1854(a)(3).

211 Jennifer Van Trump, Protecting the 'Rainforests of the Sea': Creating the Northwestern Hawaiian Islands Coral Reef Ecosystem Reserve By Executive Order, 11 PENN STATE ENVT'L L. R. 273, 284 (2003).

212 16 U.S.C. § 1851(a)(1).

213 16 U.S.C. § 1851(a)(2).

214 16 U.S.C. § 1802(28).

215 16 U.S.C. § 1802(5).

In: Marine Protected Areas
Editor: Floyd B. Mayr
ISBN: 978-1-60741-549-7

Chapter 3

Framework for the National System of Marine Protected Areas of the United States of America

National Marine Protected Areas Center

Executive Summary

Increasing impacts on the world's oceans from coastal and offshore development, overfishing, a changing climate, natural events, and other sources are straining the health of marine ecosystems and the Great Lakes. Impacts to these intricately balanced environments include declining fish populations, degradation of coral reefs and other vital habitats, threats to rare or endangered species, and loss of artifacts and resources that represent the diverse cultural heritage of the United States. The effects of these losses are significant and jeopardize the social and economic fabric of the nation.

In the United States and around the world, marine protected areas (MPAs) are increasingly recognized as an important and promising management tool for mitigating or buffering some of these impacts. When used effectively and as a part of a broader ecosystem-based approach to management, MPAs can help to restore and maintain healthy marine and Great Lakes environments by contributing to the overall protection of critical marine habitats and resources. In this way, effective MPAs also can offer social and economic opportunities for current and future generations, such as tourism, biotechnology, fishing, education, and scientific research.

MPAs are designated and managed at all levels of government by a variety of agencies including parks, fisheries, wildlife, natural resource and historic resource departments, among others. U.S. MPAs have been established by well over 100 legal authorities, with some federal and state agencies managing more than one MPA program, each with its own legal purpose. There are approximately 1,700 existing MPAs in the United States that have been established by federal, state, territorial, and local governments to protect and conserve the nation's rich natural and cultural marine heritage and sustainable production resources. These MPAs have been designated to achieve a myriad of conservation objectives, ranging from conservation of biodiversity hotspots, to preservation of sunken historic vessels, to protection of spawning aggregations important to commercial and recreational fisheries. Similarly, the level of protection provided by these MPAs ranges from fully protected or no-take marine reserves to sites allowing multiple uses, including fishing, recreational, and industrial uses.

Recognizing the significant role that U.S. MPAs play in conserving marine heritage and sustainable use, and the lack of a national institution for comprehensive MPA planning, coordination, and support, Presidential Executive Order 13158 of May 26, 2000 (Order), found in Appendix D of this document, calls for the development of a National System of Marine Protected Areas (national system). The Order clearly calls for a national and not a federal system, and requires collaboration not only with other federal agencies, but also with coastal states and territories, tribes, Regional Fishery Management Councils, and other entities, as appropriate, including the MPA Federal Advisory Committee. The Order further specifies that the national system be scientifically based, comprehensive, and represent the nation's diverse marine ecosystems and natural and cultural resources.

To provide a blueprint for building the National System of MPAs,[1] the Order calls for the development of a framework for a National System of MPAs and directs the establishment of a National MPA Center (MPA Center) within the National Oceanic and Atmospheric Administration (NOAA) to lead the system's development and implementation. This final *Framework for the National System of MPAs of the United States of America* (Framework) is the result of a multi-year development effort. The first draft Framework received over 11,000 comment submissions (composed of comments from 100 individual commenters and a petition from nearly 11,000 people) during its September 2006 to February 2007 public comment period. A second draft addressing these comments was published for public comment from March-May 2008, and received 34 public comment submissions. The MPA Federal Advisory Committee also provided two sets of recommendations on the Framework that have contributed significantly to its final form.

The Framework recognizes that U.S. MPA programs can achieve more efficient, effective conservation of the nation's important natural and cultural resources by working together rather than separately, and that many solutions require collaboration across programs with their own individual mandates, levels of government, and even international boundaries. It proposes a national system that is, initially, an assemblage of existing MPA sites, systems, and networks established and managed by federal, state, territorial, commonwealth, tribal, or local governments, acknowledging and building upon the contributions of these foundation programs. In addition, the Framework outlines collaborative, transparent processes for MPA programs at all levels of government to work together at regional, national, and international levels and with public participation to achieve common conservation objectives through comprehensive MPA planning; identification of enhanced or new MPAs that may be needed; and support for improved MPA science, stewardship, and effectiveness.

The Framework Outlines the Following Key Components of the National System

- A set of overarching national system goals and priority conservation objectives.
- MPA eligibility criteria and other key definitions.

- A nomination process for existing MPAs to be included in the national system that provides opportunities for public input.
- A science-based, public process for identifying conservation gaps in the national system.
- A process for improving regional and ecosystem-based coordination of MPAs by:
    - creating new or strengthening existing regional forums for MPA coordination;
    - identifying and catalyzing action to address shared priorities for improving MPA science, stewardship, and effectiveness; and
    - developing collaborative, ecosystem-based MPA planning to identify and recommend MPAs for inclusion in the new national system.

- Mechanisms for national and international coordination.
- Implementation guidance regarding federal agency responsibilities to avoid harm to resources protected by the National System of MPAs.
- Mechanisms for monitoring, evaluating, and reporting on national system progress and priorities.

Through collaborative efforts among U.S. MPA programs and stakeholders, the national system can achieve the Order's goal of enhancing the comprehensive conservation of the nation's natural and cultural marine heritage and the ecologically and economically sustainable use of the marine environment for present and future generations.

Marine Protected Area – *Any area of the marine environment that has been reserved by federal, state, territorial, tribal, or local laws or regulations to provide lasting protection for part or all of the natural and cultural resources therein. (Executive Order 13158)*

National System of MPAs – *The group of MPA sites, networks, and systems established and managed by federal, state, tribal, and/or local governments that collectively enhance conservation of the nation's natural and cultural marine heritage, and represent its diverse ecosystems and resources. National system MPAs work together at the regional and national levels to achieve common objectives for conserving the nation's important natural and cultural resources.*

II. Introduction

A. Background

With the world's largest Exclusive Economic Zone (Figure 1), the coastal, marine, and Great Lakes waters of the **United States**[2] support an incredible diversity and wealth of life. These waters also play host to untold special places that represent our rich cultural heritage and connections to the sea. In the same way, myriad human uses, livelihoods, and other activities take place in the marine and coastal environment, benefitting from and relying upon the sustained health of our nation's vast natural and cultural heritage.

As human populations grow and use of marine resources increases, so do the pressures and stresses exerted on these intricately balanced ecosystems. Ensuring the long-term health of these ecosystems and the sustained benefits on which humans depend requires comprehensive management approaches. In the United States and many other countries around the world, marine protected areas (MPAs) are increasingly recognized and used as important tools for the conservation and sustainable use of marine resources and as an important component of a comprehensive management approach.

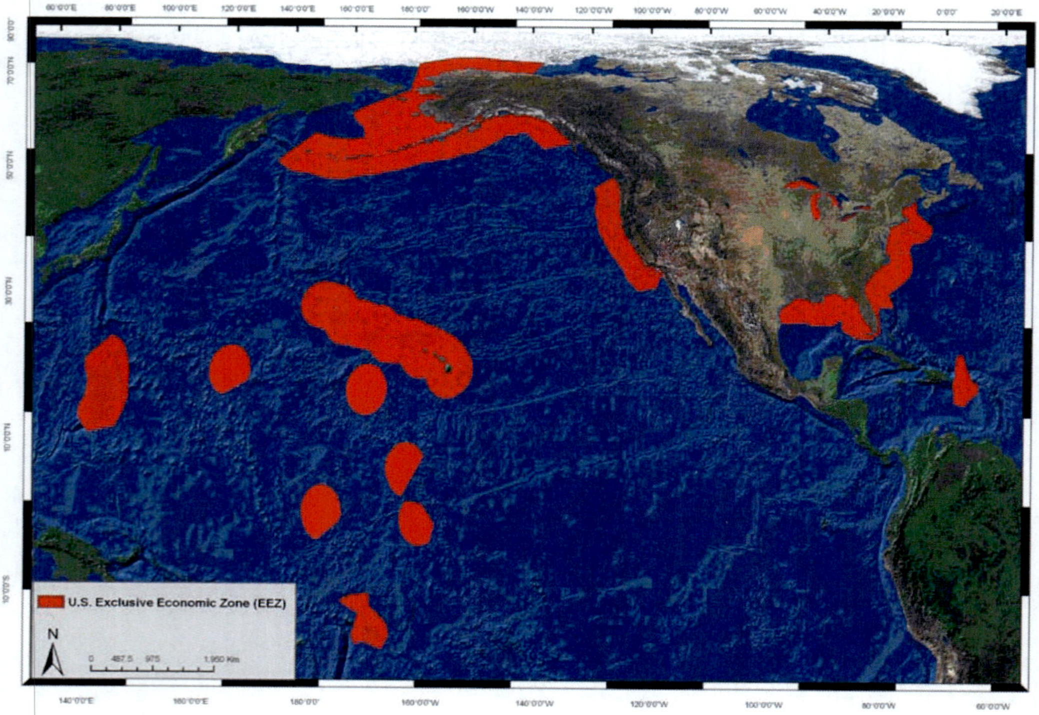

Figure 1. U.S. Exclusive Economic Zone

"Based on evidence from existing marine area closures in both temperate and tropical regions, marine reserves and protected areas will be effective tools for addressing conservation needs as part of integrated coastal and marine area management."

"MPAs, areas designated for special protection to enhance the management of marine resources, show promise as components of an ecosystem-based approach for conserving the ocean's living assets."

"Integration of management across the array of federal and state agencies will be needed to develop a national system of MPAs that effectively and efficiently conserves marine resources and provides equitable representation for the diversity of groups with interests in the sea."

Committee on the Evaluation, Design, and Monitoring of Marine Reserves and Protected Areas in the United States, Ocean Studies Board, Commission on Geosciences, Environment, and Resources, National Research Council, /Marine Protected Areas: Tools for Sustaining Ocean Ecosystems./ Washington, D.C.: National Academy Press, 2001.

Recognizing the expanding role and importance of MPAs in the United States, Presidential Executive Order 13158 of May 26, 2000 (Order) directs the Department of Commerce (DOC) and the Department of the Interior (DOI), in consultation with other federal agencies,[3] to develop a **National System of Marine Protected Areas** (national system).

The Order specifies that this is to be a *national* and not a *federal* system and requires consultation with all **states** (this includes U.S. states, territories, and commonwealths as defined in the Glossary, Section VI) that contain portions of the marine and Great Lakes environment; tribes; Regional Fishery Management Councils (FMCs); and other entities, as appropriate, including the Marine Protected Areas Federal Advisory Committee (MPA FAC) established by the Department of Commerce under the Order. The Order further specifies that the national system be scientifically based and comprehensive, and that it represent the diverse marine **ecosystems** of the United States and the nation's **natural** and **cultural resources.**

To provide a roadmap for building the national system, the Order calls for the development of a framework for a National System of MPAs and establishes the National MPA Center (MPA Center) within DOC's National Oceanic and Atmospheric Administration (NOAA) to develop the system and coordinate its subsequent implementation. This *Framework for the National System of Marine Protected Areas of the United States of America* (Framework) outlines collaborative processes for building this assemblage of existing MPA sites, networks, and systems established and managed by federal, state, tribal, or **local governments** and for collectively working together at the **regional** and national levels to achieve common objectives for conserving the nation's important natural and cultural resources.

For the purposes of the national system, the term **"marine protected area"** (MPA) is defined by the Order as, "Any **area** of the **marine environment** that has been **reserved** by Federal, State, territorial, tribal, or **local** laws or regulations to provide **lasting protection** for part or all of the natural and cultural resources therein." The term MPA, as defined and further clarified and used in this document, is not synonymous with or limited to "no-take reserves" or "marine reserves." The term MPA used here denotes an array of levels of protection and conservation purposes, from areas that allow multiple-use activities to areas that restrict take and/or access. To meet the nation's goals for conserving **natural heritage**

and cultural heritage and achieving **sustainable** production of resources found in the coastal and marine environments, the national system must include an approach to balancing types and levels of MPA protections that is science-based and **stakeholder** informed. The national system is intended to be inclusive of MPAs across the spectrum of levels of protection, from multiple-use to no-take, recognizing that existing MPAs across this spectrum offer different values to the national system that can help meet its goals and objectives.

While MPAs are an important tool for marine conservation, other types of management approaches are employed to address marine conservation objectives while allowing other appropriate uses and activities in the marine environment to take place in an economically and environmentally sustainable manner. Like other tools, MPAs should be carefully designed and implemented to meet specific conservation goals. Efforts to develop the national system must be both coordinated and integrated within the larger, evolving ecosystem-based approach to managing marine resources.

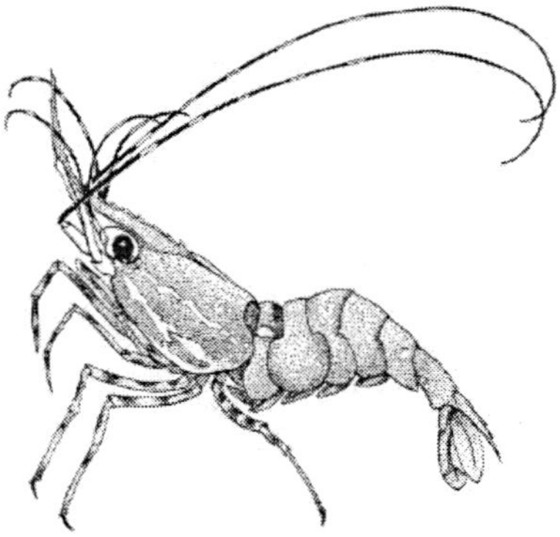

Neither the national system nor the Order establish any new legal authorities to designate, manage, or change MPAs, nor do they alter any existing federal, state, local, or tribal MPA laws or programs. Each MPA or program that participates in the national system will continue to be independently managed by its respective entity or entities, as will any new sites that eventually may be established by those authorities. The national system is intended to support, not interfere with, agencies' independent exercises of their own existing authorities. The national system is therefore envisioned as a "system of sites and systems" that will be developed to achieve conservation and management objectives that could not be accomplished by individual MPAs or MPA programs working independently.

Furthermore, the *requirements* outlined in the Order, which provide the legal authority for establishing the national system, apply only to the actions of federal agencies. The Order does not direct the actions of states or tribes, or alter any existing state, local, or tribal authorities or treaties regarding the establishment or management of MPAs or marine resources under their jurisdiction. Finally, nothing in this document is to be construed as altering existing authorities regarding the establishment of federal MPAs in areas of the marine environment subject to the jurisdiction and control of states, tribes, or local governments.

While the Order's requirements apply only to federal agencies, the full and ongoing participation of state, tribal, and local governments is critical to an effective national system. MPAs are designated and managed at all levels of government by a variety of agencies including parks, fisheries, wildlife, and natural resource and historic resource departments, among others. U.S. MPAs have been established by over 100 legal authorities, with some federal and state agencies managing more than one MPA program, each with its own legal purpose. Given the importance of the marine resources they manage and their wealth of experience in doing so, building and implementing the national system in partnership with state, tribal, and local governments is a major emphasis of the Framework. A full description of the range of existing U.S. MPA programs, federal MPA initiatives and tribal and international efforts can be found in Appendix B of this document. In light of this breadth of existing U.S. MPA responsibilities, the Order recognizes the need and calls for a national, rather than federal, system of MPAs with a geographic scope that spans the U.S. waters of the Pacific Ocean, including the Bering Sea; Atlantic Ocean, including the Gulf of Mexico and Caribbean Sea; Arctic Ocean; and the Great Lakes.

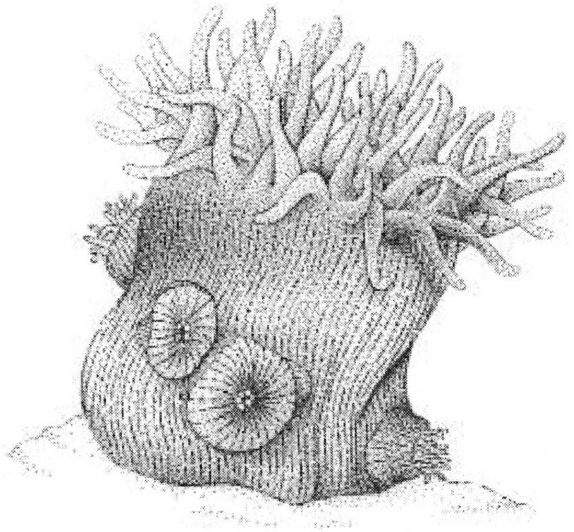

By establishing an effective structure for working together, the national system will help to increase the efficient protection of important marine resources; contribute to the nation's overall social and economic health; support government agency cooperation and integration; and improve the public's access to scientific information and decision making about the nation's marine resources. It affords all system members the protections of Section 5 of the Executive Order, which requires federal agencies to avoid harm to the natural and cultural resources protected by MPAs within the national system, to the extent permitted by law and to the maximum extent practicable. The collaborative efforts of the national system are also intended to benefit the participating federal, state, tribal, and local government partners through the identification of shared priorities for improving MPA effectiveness and the development of partnerships to provide assistance in meeting those needs. Finally, the national system provides a foundation for cooperation with other countries to conserve resources of common concern.

B. Developing the Framework

In developing this Framework, the MPA Center engaged the nation in a multi-year dialogue to ensure that the national system represents the nation's interests in the conservation and sustainable use of its natural and cultural marine resources. The MPA Center continues to work with and solicit input from federal, state, tribal, and local government partners, FMCs, stakeholder groups, and the general public about their perspectives on the national system.

Recommendations and comments from the MPA FAC, states, tribes, federal agencies, FMC representatives, and non-governmental stakeholders have provided the foundation of viewpoints and information on which this document is constructed. Moreover, many of the core concepts presented in this document stem directly from the recommendation documents and reports submitted by the MPA FAC and states.

The MPA Center led a broad and inclusive public scoping process to develop the initial draft Framework starting in 2005, and conducted general discussions about the purpose of the national system as early as 2001. Specific recommendations during the scoping process were sought and received from the MPA FAC, composed of 30 individual members of the public representing the range of the nation's MPA stakeholders and geographic areas; an MPA State Advisory Group convened by the Coastal States Organization and the MPA Center; and the Federal Interagency MPA Working Group, which provides ongoing, coordinated advice from federal agencies on the implementation of the Order. A full description of the MPA FAC can be found in Appendix B and a list of the MPA FAC members and the Federal Interagency MPA Working Group representatives can be found in Appendix E. The MPA Center also held a series of five regional public dialogue meetings around the country to provide stakeholders with an opportunity to include their input and advice and three regional state workshops to solicit their views. Comments and recommendations received during the scoping process were reviewed and considered in the development of the initial Draft Framework and copies of these and other related materials can be found at http://www.mpa.gov.

The initial Draft Framework was available for public comment between September 2006 and February 2007. The MPA Center received over 11,000 comment submissions comprised of approximately 100 comments from individual commenters and a petition from nearly 11,000 people requesting the development of a nation-wide system of fully protected or "no-take" reserves. In addition, in April and October 2007, the MPA Center solicited and received additional advice and comments from the MPA FAC about options for revising the Framework.

The Revised Draft Framework was made available for public comment from March 15, 2008, through May 16, 2008. The MPA Center received 34 comment submissions during this comment period. During both comment periods, comments were received from state government agencies, industry and conservation organizations, tribal groups, various advisory bodies, and members of the public. In developing this final Framework, the MPA Center considered all comments received during both comment periods as well as the recommendations of the MPA FAC. With the publication of this final Framework, the MPA Center will now initiate implementation of the national system. Plans and guidance documents outlining next steps in the implementation process will be posted at http://www.mpa.gov.

C. Benefits of an Effective National System

The national system offers numerous benefits above and beyond the benefits realized by participating MPA sites and programs individually. These benefits would accrue to the nation as a whole, as well as at regional and local levels. Benefits would extend across the full spectrum of users and stakeholders, including both consumptive and non-consumptive users. The following list reflects some of the potential benefits from the creation and effective management of the national system.[4]

Enhanced Conservation

- Representativeness – The national system will significantly boost ongoing efforts to preserve the natural and cultural heritage of the United States by ensuring that the diverse characteristics of the natural and social environment of the nation's seas are conserved for future generations in a systematic way. The representation of all ecosystem or habitat types in all the nation's marine regions, which includes the Great Lakes, within a single system will help ensure that the full complement of biodiversity and valued areas will be protected.

- Connectivity – The national system provides an opportunity to identify and establish networks of MPAs that are ecologically connected. An ecological network of MPAs is a set of discrete MPAs within a region that is functionally connected through dispersal of reproductive stages (eggs, larvae, spores, etc.) or movement of juveniles and adults. Properly designed and located, these networks can enhance linkages between sources and sinks for many marine organisms, which may be essential for some local populations to persist—an increasingly serious challenge in a rapidly changing environment. Planning at the national and regional scales provides an opportunity to address connectivity for many different marine organisms at different spatial scales.

- Enhanced Stewardship – The national system can help protect MPAs against the harmful effects of onsite or offsite activities through enhanced regional coordination, public awareness, site management capacity, recognition of these MPAs as important conservation areas, and application of the protective measures in Section 5 of the Executive Order.

Social and Economic Benefits

- Increased Visitation – The establishment and recognition of the national system could be an incentive for increased tourism and visitation of some MPAs, as well as an increase in visitation and enjoyment of areas system-wide, providing for uses such as recreational fishing, diving, whale watching, and swimming.

- Sustained Fisheries – One goal of the national system is supporting sustainable production of harvested marine resources. Improved regional coordination and support for management, using MPAs where appropriate, could lead to enhanced fishing opportunities for both commercial and recreational fishermen as a result of species recovery, spillover and seeding effects, habitat protection, conservation of old-growth age structure and genetic diversity, establishment of reference sites to examine the regional effects of fishing, and better information on access opportunities.

- Maintained Coastal Community Identity – Creation of the national system could help foster social stability by helping to maintain cultural heritage and economic viability.

- Non-extractive Uses – Establishment of the national system could create additional system-wide non-consumptive benefits, such as aesthetic, bequest, and spiritual values; opportunities for viewing and photographing marine wildlife; wilderness experiences; scientific research; education; and appreciation of natural resources and the importance of their management.

- Enhanced Planning for Ocean Uses – Identification of national system MPAs, as well as identification of areas important for conservation identified through a gap analysis, will help inform regional-scale planning and decision making associated with a wide range of ocean uses. This could also contribute to a more predictable regulatory environment for ocean industry.

Public Awareness, Understanding, and Education

- Increased Support for Marine Conservation – The national system recognizes the immense value of our nation's oceans and coasts and could help boost marine conservation by elevating the public profile of MPAs as a management tool. The designation of existing MPAs as part of the national system could enhance the stature of these sites within their managing entities and their local communities, as well as nationally and internationally. This designation also could build support for investment in appropriately established MPAs. Recognition of protected areas in other national or global systems (e.g., the National Estuarine Research Reserve, National Trail, and National Wilderness systems; United Nations Educational, Scientific, and Cultural Organization's World Heritage Sites; Ramsar Wetland sites) has had similar results.

- More Effective and Efficient Outreach – The national system will be an important and efficient mechanism for increased public awareness and understanding of the importance of marine resources and conservation efforts. Coordinated outreach efforts will increase the impact of outreach by individual MPAs, and could result in cost savings. Including worthy, but currently little known, sites in the national system could bring increased recognition and visibility to these areas.

- Promotion of Cultural Heritage – Participation in the national system elevates and enhances the recognition of and appreciation for the cultural heritage value of MPA sites.

- Enhanced Educational Opportunities – The creation of the national system will present enhanced opportunities for natural and cultural heritage education. This could include onsite education and interpretation, as well as classroom and web-based resources. The national system will be a valuable tool for educating students and visitors about the nation's diverse marine and coastal ecosystems and cultural resources.

- Enhanced Research Opportunities – The national system will provide scientists and managers more opportunities to understand the dynamics of marine ecosystems and human interactions with them under different management regimes.

Enhanced Coordination and Strategic Direction

- Shared National System Conservation Objectives – The national system will focus on specified priority objectives (see Section III (B)). By providing a focus for national and regional conservation efforts, these shared objectives will help build consensus about priority conservation actions, and ultimately increase the effectiveness of the diverse conservation efforts of federal agencies, states, tribes and non-governmental partners.

- Improved Gap Analysis and Planning – The formation of the national system will help highlight gaps in protection of important places for which MPAs might be considered to meet priority conservation objectives. This will inform future planning efforts to create MPAs to fill the identified gaps.

- Enhanced Interagency Cooperation – The creation of the national system will provide an unprecedented venue and catalyst for increased cooperation among the diverse entities across all levels of government with management authority for the different types of MPAs that comprise the national system. The existence of national system MPAs in the same region is intended to stimulate cooperative efforts in planning, research and monitoring, sharing of equipment and personnel, enforcement efforts, and educational campaigns.

- Enhanced Regional Coordination – The establishment or enhancement of regional MPA coordination forums via the national system offers an opportunity for managing entities and stakeholders to look beyond their individual jurisdictions, mandates, and interests, and consider regional and/or ecosystem-based approaches to MPA planning.

- Enhanced International Coordination – The national system will facilitate the identification of opportunities to improve linkages with, and provide technical assistance to, international marine protected area programs, to enhance cooperative conservation across international boundaries.

Natural Heritage: *The nation's biological communities, habitats, ecosystems, and processes and the ecological services, uses, and values they provide to present and future generations.*

Cultural Heritage: *The cultural resources that reflect the nation's maritime history and traditional cultural connections to the sea, as well as the uses and values they provide to present and future generations.*

Sustainable Production: *The nation's renewable living resources and their habitats (including, but not limited to, spawning, mating, and nursery grounds, and areas established to minimize incidental bycatch of species) and the social, cultural, and economic values and services they provide to present and future generations.*

III. Defining the National System of MPAs

A. National System Purpose

The purpose of the national system is to support the effective stewardship, conservation, restoration, sustainable use, and public understanding and appreciation of the nation's significant natural and cultural marine heritage and sustainable production marine resources, with due consideration of the interests of and implications for all who use, benefit from, and care about our marine environment.

B. National System Goals and Priority Conservation Objectives

The national system's goals and objectives are designed to address the requirements of the Order to develop a comprehensive National System of MPAs representing diverse United States marine ecosystems and the nation's natural and cultural resources. These goals, which are all of equal importance, have been designed with input and recommendations of the MPA FAC and other stakeholders to meet the purpose of the national system relative to the conservation of the nation's natural heritage, cultural heritage, and sustainable production marine resources (Table 1).

These goals and associated priority conservation objectives are intended to guide the development of the comprehensive national system, including identification of both existing MPAs to be included and conservation gaps which might be addressed through the establishment of MPAs. The national system as a whole will work collectively to achieve these goals and objectives. It is not expected that any individual MPA, MPA program, or system should address all goals or objectives. Measuring progress toward the attainment of these goals is addressed in Section V(C).

Prioritization of Conservation Objectives

Given the magnitude of the task of building a comprehensive national system, the MPA Center will follow a gradual implementation process based on the iterative achievement of the prioritized conservation objectives as outlined in the table below. In this way, building the national system will begin with a focus on a subset of the highest-priority (near-term) objectives for each goal and as completed will move on to the next highest-priority conservation objectives for each goal.

The conservation objectives listed below were prioritized by the MPA FAC and the MPA Center for near-term, mid-term, and long-term implementation based on:

- the availability of existing scientific or other data necessary to achieve the objective;
- the importance of the objective, i.e., its relative urgency and significance as compared to the other objectives; and
- the effort necessary to achieve the objective, in this case the ability to complete the nomination of existing areas and the identification of conservation gaps relative to the objective(s).

Achievement or completion of each conservation objective will include the following activities:

- identification of existing MPAs that contribute to that objective and nomination of those MPAs by managing entities to the national system, and
- identification of associated conservation gaps in the national system.

Table 1. National System Goals and Priority Conservation Objectives

Goal 1: For Natural Heritage Marine Resources – Advance comprehensive conservation and management of the nation's biological communities, habitats, ecosystems, and processes and the ecological services, uses, and values they provide to present and future generations through ecosystem-based MPA approaches.	
Priority Conservation Objectives for Goal 1 – Conserve and manage:	
Key reproduction areas and nursery grounds	**Near Term**
Key biogenic habitats	
Areas of high species and/or habitat diversity	
Ecologically important geological features and enduring/recurring oceanographic features	
Critical habitat of threatened and endangered species	
Unique or rare species, habitats, and associated communities	**Mid Term**
Key areas for migratory species	
Linked areas important to life histories	**Long Term**
Key areas that provide compatible opportunities for education and research	
Goal 2: For Cultural Heritage Marine Resources – Advance comprehensive conservation and management of cultural resources that reflect the nation's maritime history and traditional cultural connections to the sea, as well as the uses and values they provide to present and future generations through ecosystem-based MPA approaches.	
Priority Conservation Objectives for Goal 2 – Conserve and manage:	
Key cultural and historic resources listed on the National Register of Historic Places (NRHP)	**Near Term**
Key cultural and historic resources determined eligible for the NRHP or listed on a State Register	
Key cultural sites that are paramount to a culture's identity and/or survival	
Key cultural and historic sites that may be threatened	**Mid Term**
Key cultural and historic sites that can be utilized for heritage tourism	
Key cultural and historic sites that are underrepresented	**Long Term**
Goal 3: For Sustainable Production Marine Resources – Advance comprehensive conservation and management of the nation's renewable living resources and their habitats (including, but not limited to, spawning, mating, and nursery grounds and areas established to minimize bycatch of species) and the social, cultural, and economic values and services they provide to present and future generations through ecosystem-based MPA approaches.	
Priority Conservation Objectives for Goal 3 – Conserve and manage:	
Key reproduction areas, including larval sources and nursery grounds	**Near Term**
Key areas that sustain or restore high-priority fishing grounds	
Key areas for maintaining natural age/sex structure of important harvestable species	**Mid Term**
Key foraging grounds	
Key areas that mitigate the impacts of bycatch	
Key areas that provide compatible opportunities for education and research	**Long Term**

Priority conservation objectives should be considered together and at the regional scale, recognizing that implementation of the priority conservation objectives may not occur simultaneously and that conservation gaps in some areas may be addressed by MPAs, some other management tool, or a combination of tools, as appropriate. Specific processes for each of these activities are described in later sections of this document. Nonetheless, in practical terms, it is unlikely that all objectives within the same timeframe designation (e.g., near-term) will be able to be addressed simultaneously due to varying complexity of implementation and available staffing and funding resources.

To ensure that partners and stakeholders are kept informed of the status of building the national system, the MPA Center will publish, on an as-needed and sequential basis, "priorities announcements" that list the specific subsets of the near-term, mid-term, and long-term national system conservation objectives for each goal as targets for building the national system.

C. National System Design and Implementation Principles

The following principles are intended to guide the decisions and actions of managing entities and stakeholders in building and implementing an effective national system. These principles have been adapted from recommendations of the MPA FAC and the World Commission on Protected Areas/International Union for Conservation of Nature (WCPA/IUCN) report, "Establishing networks of marine protected areas: A guide for developing national and regional capacity for building MPA networks" (WCPA/IUCN, 2007).

National System Design Principles

Design principles will be used to guide the development of the national system, including the identification of priority conservation gaps in the national system (Section IV (D)) and regional MPA planning (Section V (A) (2)).

- Prioritized resource conservation targets – Focus first on conservation objectives that are of highest priority based on significance and urgency, availability of existing scientific and other data, and ability of the managing entity(ies) to act on objectives in the near-term.
- Representativeness –
  - *Geographically representative* – represents the range of geographic regions of the nation.
  - *Ecologically representative* – represents the range of marine and coastal biological diversity (from genes to species to habitats to ecosystems) and associated physical environments within the region or nation.

- *Culturally and/or historically representative* – represents the range of cultural and/or historic resources and values of a particular ecosystem or region or the nation.

- *Levels of government* – includes areas managed by federal, state, tribal, and local governments and communities.

- Replication – Includes multiple sites to ensure continued representation in the face of harmful impacts.

- Precautionary design – Decisions are based on the best information currently available from natural science, social science, customary and local knowledge, and other sources. Where information is limited, decisions should reflect a precautionary approach.

- Resilience – Designed to maintain ecosystems' natural states and to absorb shocks, particularly in the face of large-scale and long-term changes (such as climate change).

- Viability – Inclusion of self-sustaining, geographically dispersed component sites of sufficient extent to ensure population persistence through natural cycles of variation.

- Connectivity – Maximize and enhance the linkages among individual MPAs, groups of MPAs within a given eco-region, or MPA networks in the same and/or different regions.

National System Planning and Implementation Principles

Planning and implementation principles that will guide national system efforts are discussed further under Section V, "Implementing the National System," including regional coordination and MPA planning.

- Cooperation and coordination – Fosters cooperation and coordination among federal, state, tribal, local, and other management entities to reduce administrative costs, promote efficiency, and effectively utilize existing management infrastructure.

- National scope, ecosystem and regional scale – Embraces regional and ecosystem approaches to planning, participation, and implementation. Provides a mechanism for coordinating across regions, nationally, and where appropriate, internationally.

- Adaptive management – Employs a systematic process for continually improving national system management policies and practices by learning from the outcomes of operational programs.

- Monitoring and assessment – Promotes sound monitoring and evaluation at the site and system levels to assess management effectiveness, relying on established evaluation processes and methodologies, where possible.
- Compliance and enforcement – Promotes effective compliance with and enforcement of MPA regulations through design recommendations for MPAs and networks, capacity building, public education, and other mechanisms.
- Balanced stakeholder involvement – Provides meaningful opportunities for input from and participation by the nation's MPA stakeholders, including the general public.
- Active outreach and education – Raises awareness and understanding of MPAs and stewardship of marine resources.
- On-site and off-site influences and impacts – Recognizes and seeks appropriate mechanisms to address both on-site and off-site influences, including impacts to coastal and marine resources from land-based activities.
- Respecting local and indigenous values – Considers and addresses local values, including those of indigenous cultures.
- Appropriate access and compatible uses – Provides opportunities for appropriate access to and/or compatible use of marine resources consistent with conservation goals and objectives.

D. MPA Eligibility Criteria

To be eligible for nomination to the national system, existing MPAs must meet three (four for cultural sites) criteria, shown in Figure 2 and described in more detail below:

1. Meet the definitional criteria of an MPA, including each of its key terms (see definitions in Table 2) – area, marine environment, reserved, lasting, and protection.
2. Have a management plan.
3. Support at least one priority goal and conservation objective of the national system.
4. Cultural heritage MPAs also must conform to criteria for including sites on the National Register of Historic Places.

Additional sites not currently meeting the management plan criterion can be evaluated for eligibility to be nominated to the system on a case-by-case basis based on their ability to fill gaps in national system coverage of the priority conservation objectives and design principles described in Sections III (B) and (C), respectively. To the extent practicable, the MPA Center intends to assist otherwise qualified sites that do not meet the management plan criterion to develop or strengthen their management plans.

Table 2. Definition of Key Terms for the Purposes of the National System

Key Term	Definition
Area	Must have legally defined geographical boundaries, and may be of any size, except that the site must be a subset of the United States federal, state, local, or tribal marine environment in which it is located. Application of this criterion would exclude, for example, generic broad-based resource management authorities without specific locations and areas whose boundaries change over time based on species presence. The area must be one over which the United States has jurisdiction, consistent with international law.
Marine environment	Must be: (a) ocean or coastal waters (note: coastal waters may include intertidal areas, bays or estuaries); (b) an area of the Great Lakes or their connecting wat-ers; (c) an area of submerged lands under ocean or coastal waters or the Great Lakes or their connecting waters; or (d) a combination of the above. The term ''intertidal'' is understood to mean the shore zone between the mean low water and mean high water marks. An MPA may be a marine component part of a lar-ger site that includes uplands; however, the terrestrial portion is not considered an MPA. For mapping purposes, an MPA may show an associated terrestrial protected area. For purposes of the national system, NOAA and DOI intend to use the following definition for the term ''estuary'': ''part of a river or stream or other body of water having unimpaired connection with the open sea, where the sea water is measurably diluted with fresh water derived from land drainage, and extending upstream to where ocean-derived salts measure less than 0.5 parts per thousand during the period of average annual low flow.'' Application of this criterion would exclude, for example, strictly freshwater sites outside the Great Lakes region that contain marine species at certain seasons or life history stages unless that site is a component of a larger, multi-unit MPA. Upon request, the agencies will work with individual federal, state, and tribal MPAs and programs to examine unique conditions that may affect applicability of the term ''estuary'' or "coastal waters" for sites that have national or regional significance or representativeness. Estuarine-like sites on tributaries of the Great Lakes will be considered for inclusion if they are located within the eight-digit U.S. Geological Survey cataloging unit adjacent to a Great Lake or its connecting waters
Reserved	Must be established by and currently subject to federal, state, local, or tribal law or regulation. Application of this criterion would exclude, for example, privately created or maintained marine sites.
Lasting	For natural heritage and cultural heritage MPAs, the site's authority must clearly state its intent to provide permanent protection. This definition recognizes that subsequent.
	to establishment, MPA designation and level of protection may cha-nge for various reasons, including natural disasters that may destroy or alter res-ources or changes in societal values. Should any of these changes occur, the status of the MPA relative to the national system could be re-evaluated. Sites and/or protections that must have a specific legislative or other administrative action to be decommissioned shall be considered to have been established with the intent to provide permanent protection. This would include, for example, sites that have a requirement for periodic renewal contingent on evaluation of effectiveness, with no specified expiration date. For sustainable production MPAs, the site must be established with the intent at the time of designation to provide, at a minimum, the duration of protection necessary to achieve the mandated long-term sustainable production objectives for which the site was established. For all MPAs, the site must provide the same level and type of protection at a fixed location and fixed and regular period of any duration during a year

Protection	Must have existing laws or regulations that are designed and applied to afford the site with increased protection for part or all of the natural and submerged cultural resources therein for the purpose of maintaining or enhancing the lasting conservation of these resources, beyond any general protections that apply outside the site. Application of this criterion would exclude restricted areas that are established for purposes other than conservation. The term would not include, for example, areas closed for navigational safety, areas closed to safeguard modern human-made structures (e.g., submarine cable no-anchor zones), polluted shellfish-bed closure areas, areas closed to avoid fishing gear conflicts, and areas subject to area-based regulations that are established solely to limit fisheries by quota management or to facilitate enforcement.

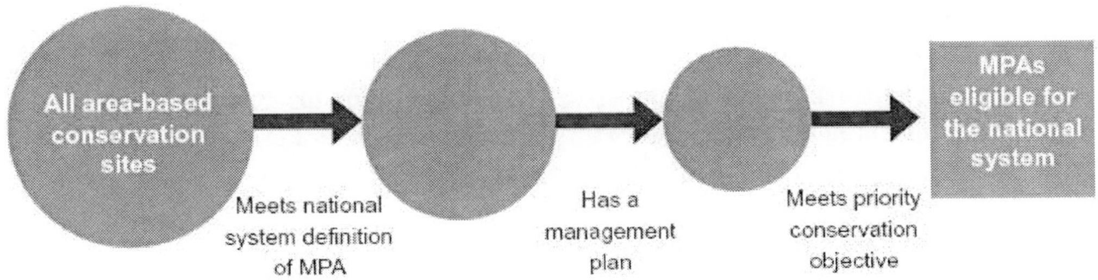

Figure 2. Eligibility Criteria for the National System

(i) Definition of MPA and its Key Terms

With the goal of standardizing the term "marine protected area" for the purposes of the national system, the Order defines an "MPA" as "[a]ny **area** of the **marine environment** that has been **reserved** by Federal, State, territorial, tribal or local laws or regulations to provide **lasting protection** for part or all of the natural and cultural resources therein."

Without further clarification, the key terms of "area," "marine environment," "reserved," "lasting," and "protection" found in the MPA definition are subject to a range of interpretations and lead to an uncertain scope for the national system. The definitions of key terms for "MPA" listed in Table 2 were guided by recommendations from stakeholders, including the MPA FAC, the analysis of existing place-based conservation efforts, and *Federal Register* comment processes for the Draft and Revised Draft Frameworks.

(ii) Management Plan Criteria

To be eligible for nomination to the national system, an MPA must have a management plan that:

Has been developed at one of the following scales:

- a site-specific MPA management plan,
- part of a larger MPA programmatic management plan,
- component of a broader, non-MPA programmatic management plan (e.g., fishery management plan or species recovery plan), or
- a verbal or written community agreement.[5]

Includes both of the following components:

- specified conservation goals, and
- a process or requirement for monitoring and evaluation of goals.

(iii) Priority Goals and Objectives of the National System

An MPA's conservation purpose must specifically contribute to at least one of the priority goals and objectives published by the MPA Center as current conservation priorities, as described in Section III (B) above.

(iv) National Register of Historic Places Criteria

Cultural resources in the national system of MPAs can include submerged archeological resources, cultural landscapes, and structures as well as ethnographic resources with tribal or traditional cultural meaning, value, and use. Given the cultural resource management community's widespread acknowledgement of the standards developed by the National Park Service for inclusion of a cultural resource in the National Register of Historical Places (NRHP), the national system will integrate core elements of those standards into its criteria for MPAs with cultural marine resources. As such, the cultural marine resources within those MPAs must be historic and defined as at least 50 years of age, unless otherwise determined to be unique to the nation's maritime history or traditional connections to the sea as defined by the NRHP. In addition, the resources must meet the following NRHP evaluation criteria:

"The quality of significance in American history, architecture, archeology, engineering, and culture is present in districts, sites, buildings, structures, and objects that possess integrity of location, design, setting, materials, workmanship, feeling, and association, and:

a. That are associated with events that have made a significant contribution to the broad patterns of our history; or

b. That are associated with the lives of significant persons in our past; or
c. That embody the distinctive characteristics of a type, period, or method of construction, or that represent the work of a master, or that possess high artistic values, or that represent a significant and distinguishable entity whose components may lack individual distinction; or
d. That have yielded or may be likely to yield, information important in history or prehistory."

E. MPA Categories

The set of national system MPA categories listed below in Table 3 are intended to provide a limited set of user-friendly terms for communicating generally about the purpose of and level of protection for MPAs that become a part of the national system.[6] In addition, these categories will be useful for:

- partitioning the national system into manageably sized groups of comparable sites to ease identification of shared technical or other assistance;
- grouping sites based on comparable conservation objectives and levels of protection to facilitate identification of gaps in conservation; and
- providing a logical framework for organizing and monitoring how sites added to the national system contribute to the system's conservation objectives.

The MPA Center will work with the respective managing entities to determine the most appropriate category for the MPAs as they become a part of the national system. This categorization will not in any way supersede the designated name or title of the MPA, as established by law or other independent authorities.

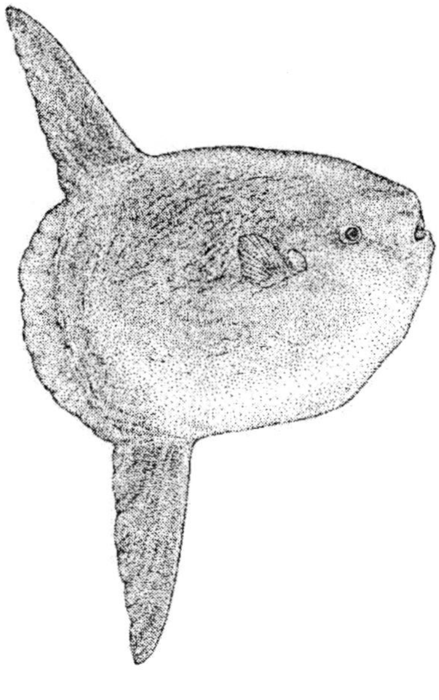

Table 3. National System MPA Categories

National System Purpose	MPA Category	Protection and Use Sub-category*	Management Goal(s)
Conserve Marine Heritage	**Marine Natural Heritage Areas**	**Natural Heritage Conservation Areas**	Conserve and manage the nation's biological communities, habitats, ecosystems, and processes and the ecological services, uses, and values they provide to present and future generations through ecosystem-based MPA approaches.
		Natural Heritage Reserve Areas	Strongly protect the nation's biological communities, habitats, ecosystems, and processes and the ecological services, uses, and values they provide to present and future generations through ecosystem-based MPA approaches.
	Marine Cultural Heritage Areas	**Cultural Heritage Conservation Areas**	Conserve and manage cultural resources that reflect the nation's maritime history and traditional cultural connections to the sea and the uses and values they provide to present and future generations through ecosystem-based MPA approaches.
		Cultural Heritage Reserve Areas	Strongly protect cultural resources that reflect the nation's maritime history and traditional cultural connections to the sea and the uses and values they provide to present and future generations through ecosystem-based MPA approaches.
	Marine Natural and Cultural Heritage Areas	**Natural and Cultural Heri-tage Conserva-tion Areas**	Management goals of marine natural heritage conservation areas and of marine cultural heritage conservation areas.
		Natural and Cul-tural Heritage Reserve Areas	Management goals of marine natural heritage reserve areas and of marine cultural reserve areas.
Sustain Marine Production	**Marine Sustain-able Product-ion Areas**	**Sustainable Production Conservation Areas**	Advance comprehensive conservation and management of the nation's renewable living resources and their habitats (including, but not limited to, spawning, mating, and nursery grounds and areas established to minimize bycatch of species) and the social, cultural, and economic values and services they provide to present and future generations through ecosystem-based MPA approaches.
		Sustainable Production Reserve Areas	Strongly protect the nation's renewable living resources and their habitats (including, but not limited to, spawning, mating, and nursery grounds and areas established to minimize bycatch of species) and the social, cultural, and economic values and services they provide to present and future generations through ecosystem-based MPA approaches.

***Conservation Areas:** Multiple uses allowed; however, uses and activities may be restricted or zoned, and access limited, as necessary to meet site management goals.

***Reserve Areas:** No extractive uses allowed, except permitted scientific and educational uses; destructive or disruptive activities limited; other uses and activities may be restricted or zoned, and access limited, as necessary to meet site management goals.

IV. BUILDING THE NATIONAL SYSTEM OF MPAS

A. Summary and Sequence

Building the national system will involve two major sets of activities:

- the identification, nomination, and inclusion of existing MPAs in the national system and on the official List of National System MPAs, and
- the identification of national system conservation gaps in protection of important marine areas that meet the national system's conservation objectives and design criteria, outlined in Sections III (B) and (D) above, with facilitation of subsequent development by the relevant establishing agencies of new MPAs and/or enhancement of existing MPAs to fill those gaps, where appropriate, outlined in Section IV (D) below.

Given the magnitude of the task of developing the national system, the MPA Center will follow an iterative process to build the system gradually over time. The pace of this process will be determined by the availability of resources to carry out the process. The sequence of the iterative process for the above two major sets of national system building activities is as follows, and shown in Figure 3 (a more thorough description of each activity can be found in subsequent subsections):

- As described in Section III (B), the MPA Center will periodically identify near-term priority conservation objectives to guide the phased development of the national system.
- As described in Section IV (B), the MPA Center will lead a nation-wide nomination process for eligible existing MPAs that contribute to the targeted conservation objectives, and include those MPAs in the national system that are successfully nominated and accepted.
- As described in Section IV (D), the MPA Center will lead a collaborative region-by-region process to identify conservation gaps relative to the targeted conservation

objectives and national system design criteria. Conservation gaps will be used to inform the development of recommendations for new MPAs through regional MPA planning described in Section V (A), and can also be used by managing entities and stakeholders to guide their efforts to establish new MPAs. It is expected that any management actions taken to fill these gaps will consider different management alternatives and the impacts of those alternatives on human uses of the areas.

- Upon completion of the nation-wide nomination process and region-by-region conservation gap identification for the targeted conservation objectives, or at such other time that resources and capabilities allow, the MPA Center will publish the next iterative set of conservation objectives to serve as targets for building the national system.

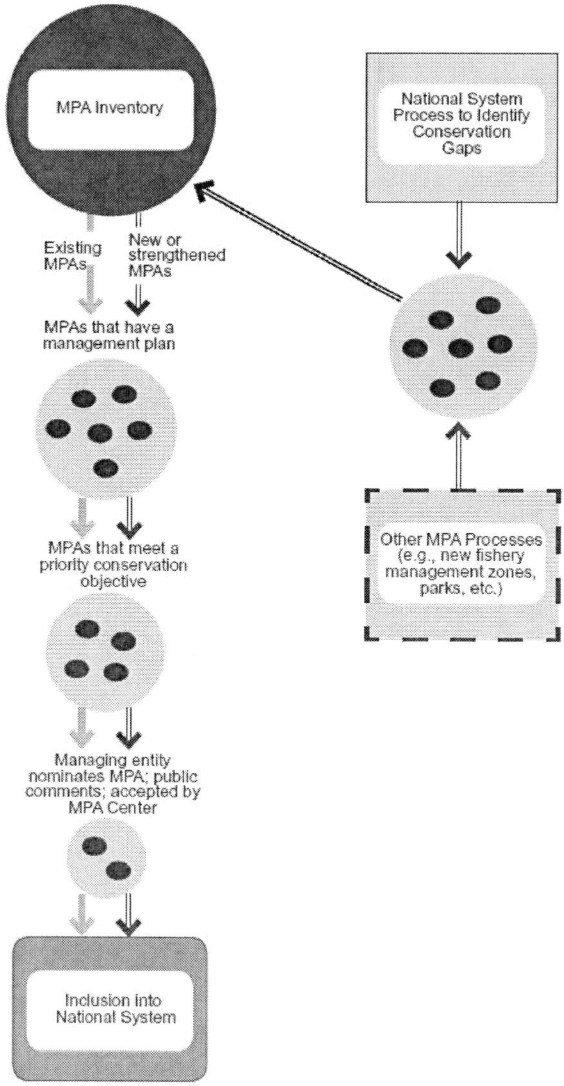

Figure 3. Building the National System of MPAs

B. Nomination Process for Existing MPAs

The process for nominating and including eligible MPAs in the national system is as follows. Nominations of existing MPAs originate with the managing entity(ies), with the MPA Center providing background information and analysis (see Figure 4 for summary):

1. The MPA Center will review sites in the United States Marine Protected Areas Inventory and identify the set of sites that meet the three (or four, for cultural sites) MPA eligibility criteria outlined in Section III (D). Information on whether sites meet criterion 3, supporting at least one priority goal and conservation objective of the national system, will be provided by the managing entity. The MPA Inventory (see http://www.mpa.gov) is a refinement of the earlier Marine Managed Areas Inventory, which was a broader collection of place-based management areas in U.S. waters.
2. The MPA Center will send the managing entity or entities[7] for those sites found to be potentially eligible a letter of invitation to nominate the site, including the rationale for eligibility.
3. The managing entity or entities will be asked to consider and nominate some or all of the identified sites for inclusion in the national system, including additional information required to evaluate site eligibility relative to meeting priority conservation objectives.
   The managing entity or entities may also provide a brief justification and nomination for: a) unsolicited sites believed to meet the requirements for entry into the national system, or b) other sites that do not appear to currently meet the management plan eligibility criterion but are deemed to be a priority for inclusion based on their ability to fill gaps in national system coverage of the priority conservation objectives and design principles.
4. The MPA Center will review the set of nominated sites to ensure that nominations are sufficiently justified.
5. The MPA Center will notify the public, via the Federal Register and other means, of the set of sites nominated for inclusion in the national system and provide the opportunity to comment on the eligibility of nominated sites (or sites that have not been nominated) relative to the eligibility criteria and any additional justification. The MPA Center will work with the managing entities to ensure adequate public involvement, including public meetings, as appropriate.
6. The MPA Center will receive, evaluate, and forward public comment to the relevant managing entity or entities, which will reaffirm or withdraw (in writing to the MPA Center) the nomination based on public comment received and any other factors deemed relevant.
7. The MPA Center will review the final determination for each nomination, consult as necessary with the managing entity or entities should there be any discrepancies, and accept mutually agreed upon MPAs into the national system.
8. MPAs that are accepted into the national system will be listed in the official List of National System MPAs (see below) comprising the national system and made available to the public via the Federal Register, the website http://www.mpa.gov, and other means.

Figure 4. Summary of Nomination Process

Where non-governmental stakeholders, including the general public, may have an interest in the nomination of certain MPAs, they are encouraged to contact the respective managing entity or entities to share their perspectives about nomination in addition to participating in the public comment process described in number 5 in this section. Similarly, where government agencies have an interest in the nomination of eligible MPAs for which they do not have management authority, they are encouraged to consult with the respective managing entity or entities.

C. The Official List of National System MPAs

1. Adding MPAs to the List and National System

Pursuant to Section 4(d) of the Order, and to ensure that managing entities, organizations, and the general public are aware of the MPAs that make up the national system, the MPA Center will maintain a List of National System MPAs. The List of National System MPAs will be the official inventory of all MPAs that have been formally included in and recognized as part of the National System of MPAs under Section IV (B), above. In addition, MPAs on the List of National System MPAs are those sites that are the subject of Section 5 of the Order, "Agency Responsibilities," as described in Section V (D) of this document. This authority does not apply to MPAs not on the List of National System MPAs.

The List will include the following information for each national system MPA:

a. name,
b. location,
c. national system MPA category,
d. priority conservation objective(s) contributed to,
e. boundaries,
f. key resources protected,
g. authorizing legislation,
h. levels and types of protection,
i. managing authority or program,
j. name of point of contact, and
k. relevant contact information.

The MPA Center will regularly publish an updated, summary version of the List of National System MPAs in the *Federal Register*, and will make it available to the public at http://www.mpa.gov or by request.

2. Modifying MPAs on the List and in the National System

Participation in the national system does not constrain the management entity from changing its management of the MPA. The management entity would still have the ability, within its own authorities and required processes, to add or reduce levels of protection, change the size of the MPA, or make other changes. Management entities would be asked to provide all significant updates to the MPA Center, but would not be required to re-nominate the site. If the MPA no longer meets the national system MPA eligibility criteria, it would be removed from the system (see Section IV (C) 3).

3. Removing MPAs from the List and National System

MPA sites or systems that have been included on the List of National System MPAs may be removed at any time by written request of the managing entity(ies) or the MPA Center for reasons including:

- the MPA ceases to exist (e.g., the legal authority or regulations expire);
- the MPA no longer meets the national system MPA eligibility criteria; or
- the managing authority requests removal.

All requests from managing entities or actions by the MPA Center to remove an MPA from the national system must be made in writing, will become part of the public record, and will be published at http://www.mpa.gov and in the *Federal Register* for comment. Upon receipt by the MPA Center of a request to remove an MPA from the national system, the managing entity(ies) and the MPA Center will enter into a dialogue on the proposal. Any comments received from the public relating to the removal of an MPA from the national system will be forwarded to the managing entity(ies) for its consideration in making its final determination to have the site removed from the national system. Upon completion of all obligations by the respective managing entity(ies), the MPA will be removed from the List of National System MPAs and all information referencing the site will be removed from national system materials and archived in the national system information on the website.

D. Identifying National System Conservation Gaps

The nation's suite of existing MPAs contributes significantly to the building of a comprehensive and representative national system. The critical next step toward achieving the national system's conservation objectives is the identification of conservation gaps: areas in the ocean and Great Lakes that meet priority conservation objectives of the national system but that are currently not adequately protected to ensure their long-term viability, as called for in Section 4 (a) of the Order. Conservation gaps identified herein can be used by existing federal, state, tribal, and local MPA managing entities and others to guide their future efforts to establish new or strengthen existing MPAs using their independent authorities and processes, or to address these gaps through other management tools. In addition, the gaps identified through this process will be used to facilitate regional planning and collaboration that may ensue as described in Section V (A).

This section outlines the process for identifying gaps in the national system. The process will be comprehensive, taking into account existing MPAs and other conservation measures currently in place. The gap analysis process will be implemented iteratively, relative to targeted specific national system conservation objectives, and on region-by-region bases as described below. Conservation gaps in the national system may exist in a number of forms and can be generally described as:

Representation gaps: where a particular habitat, ecosystem, or cultural resource type is either un-represented or underrepresented in the national system.

Ecological gaps: where important species, habitats, ecosystems, or processes fundamental to the national system's goals are not adequately protected to ensure their lasting conservation and sustainable use.

Management gaps: where the management regimes (management objectives or governance types) of MPAs in the national system do not fully provide for lasting conservation or sustainable production of a particular species, habitat, cultural resource, or ecosystem.[8]

Efforts to identify conservation gaps will include the collection and analysis of the best available scientific information and analyses, including traditional ecological knowledge, to identify important marine areas on multiple scales, coupled with an analysis of existing levels of place-based protection in those areas. The resulting gaps in protection will be identified relative to fully achieving the national system conservation objectives and design principles outlined in Sections III (B) and (C), respectively. Gap identification efforts will be focused at the regional scale, and will be collaborative, involving MPA-related and other entities at various levels of government, FMCs, and other organizations and institutions in synthesizing and analyzing existing scientific information, including traditional ecological knowledge, where available, and established conservation priorities. The effort to identify conservation gaps will include opportunities to review and comment on the process and its results by the public, the MPA FAC, relevant federal agencies, state and tribal governments, and other entities, including the National System Management Committee (Management Committee) described in Section V (B).

The MPA Center also will work with existing or incipient regional marine entities and initiatives to coordinate with their broad management efforts, as appropriate. Efforts to identify gaps will also consider and include relevant international participation and linkages. The effort aims to provide government agencies with a program-neutral opportunity for collaborative assessment and planning, while ensuring that stakeholders are both informed and involved.

The MPA Center will work with diverse partners, as appropriate, through the following processes to identify gaps in fully achieving the national system's conservation objectives:

- Publish, on an as-needed and sequential basis, subsets of the near-term, mid-term, and long-term national system conservation objectives listed in Section III (B) as iterative targets for conservation gap identification.
- On a regional basis, aggregate, map, and describe relevant and readily available existing data and analyses about important species, habitats, cultural resources, and ecosystems that could contribute to the national system goals and priority conservation objectives.
- Map and describe, by region, the location and management attributes of existing MPAs that contribute to achieving the targeted national system conservation objectives.
- Integrate spatial data on ecosystems and place-based management to identify important areas where protection is either lacking or potentially inadequate to achieve national system goals and objectives.
- Identify key stakeholders in the region and provide identified gaps and background information to the public for comment.
- Seek input on identified gaps from federal agencies, states, and tribal leaders with management authority in the corresponding region.
- Seek input on identified gaps from the Management Committee.
- Provide identified gaps, background information, and a summary of all public and Management Committee comments received to the MPA FAC for consideration and development of prioritized recommendations to DOC and DOI.

- Upon consideration of all input and recommendations, the MPA Center will publish prioritized national system conservation gaps and corresponding descriptive information for use by managing entities and stakeholders to strengthen existing MPAs or add new MPAs where needed. Information about the conservation gaps identified will be maintained on the http://www.mpa.gov website. Gap analyses will be updated periodically as resources permit.

Finally, while the publication of these identified conservation gaps is a major step toward building a comprehensive national system, significant additional evaluation of these gaps and other information will likely be needed by agencies prior to any resulting establishment of new MPAs or changes to existing MPAs' governance. Specifically, managing entities will need to work with stakeholders under the auspices of appropriate MPA authorities to: (i) evaluate these gaps; (ii) incorporate data on human uses and impacts and related societal and economic considerations; and (iii) assess management priorities to make an informed decision about appropriate next steps in response to an identified conservation gap. These steps might include the establishment of a new MPA, changes to existing MPAs, additional research, or some other alternative. Establishment of new MPAs or changes to the governance of existing MPAs must follow relevant processes under established authorities.

The MPA Center can serve as a resource to assist managing entities and stakeholders with such analyses and regional planning processes, as described in Section V (A). Similarly, identified gaps will be considered by the MPA Center and the Management Committee in prioritizing national system science and stewardship actions. The MPA Center also will report on actions taken by managing entities to address these gaps.

E. Establishing New National System MPAs

The Framework lays out the processes for identifying conservation gaps in the national system (see Section IV (D)) and developing recommendations for new or enhanced MPAs through collaborative ecosystem-based MPA planning (see Section V (A)(2)). However, neither the Order nor the Framework provides authority to designate or establish new MPAs or alter protections afforded by existing MPAs. Section 4(e) of the Order states:

The goal of the MPA Center shall be, in cooperation with the Department of the Interior, to develop a framework for a national system of MPAs, and to provide Federal, State, territorial, tribal, and local governments with the information, technologies, and strategies to support the system. This national system framework and the work of the MPA Center is intended to support, not interfere with, agencies' independent exercise of their own existing authorities.

These national system processes are intended to offer a more collaborative, systematic and comprehensive approach to MPA planning than currently exists. Recommendations for new or enhanced MPAs that stem from these processes offer entities with MPA management authority valuable guidance for taking independent or cooperative action to establish and/or manage MPAs that meet program mandates while also enhancing regional and national conservation priorities. Moreover, such processes and recommendations offer stakeholders opportunities and information with which to meaningfully engage in MPA decision making efforts.

New MPAs that may eventually be established based on these national system recommendations would subsequently be considered for inclusion in the national system pursuant to the eligibility criteria and nomination process outlined above. Stakeholder participation in the designation process for new MPAs is unchanged by the national system and occurs as specified through the required public consultation processes associated with the authorized designation process.

V. IMPLEMENTING THE NATIONAL SYSTEM OF MPAS

Over time, as MPA sites, programs, and systems are added to the national system, efforts to implement the national system – both regionally and nationally – will be initiated. A major emphasis of the MPA Center will be to facilitate and support collaborative implementation efforts with participating MPA sites and programs, subject to available resources. The timing of the implementation elements, described below, may be sequential, simultaneous, or otherwise, depending on resources available and the priorities of national system partners. Significant additional resources will be needed to realize the full potential of each element. In addition, monetary and nonmonetary incentives would greatly enhance state, tribal, and local participation in the national system, thereby increasing its conservation impact. National system implementation components, guided by the national system's design planning and implementation principles described in Section III (C), include:

- *Enhancing regional coordination and collaboration* – formalizing new and/or supporting existing regional mechanisms to provide for effective, efficient coordination and collaboration among participating MPA sites, systems, and programs.

- *Improving MPA stewardship and effectiveness* – identifying and prioritizing shared needs for improvements in MPA science, management, and stewardship at regional and national levels and catalyzing partnerships and action to address identified priorities for existing MPAs.

- *Regional MPA planning* – developing and applying the natural and social science information, decision making tools, and stakeholder engagement processes to evaluate collaboratively the conservation gaps identified in the national system and make recommendations about the need for new and/or enhanced MPAs.

- *National and international coordination* – establishing and implementing a National System Management Committee to serve to link across regions where resource conservation and MPA planning and management issues span regional boundaries and to identify and pursue international MPA linkages to the national system.

- *Evaluating national system effectiveness* – providing technical and scientific support for fostering sound monitoring and evaluation programs at the participating MPA site or system level, as well as development of a set of standards and protocols for assessing broader national system effectiveness.

- *Federal agency responsibilities to avoid harm* – providing guidance regarding Section 5 of the Order, which requires federal agencies to "avoid harm" to the natural and cultural resources protected by MPAs that become part of the national system.

- *Tracking and reporting* – maintaining the http://www.mpa.gov website and producing a biennial State of the National System report and other mechanisms for communicating national system activities, progress, and plans.

A. Enhancing Regional Coordination and Collaboration

Within the national system, effective regional coordination and collaboration are critical for sharing information and experiences, identifying common priorities and collaborative solutions for enhancing the effectiveness of existing sites, and improving planning and decision making for new MPAs. In the same way, effective regional collaboration must also include making necessary linkages to other marine management initiatives and collaboration mechanisms. For example, the federal Seamless Network initiative, the developing U.S. Integrated Ocean Observing System, coordination with the Regional Fishery Management Councils and Inter-State Fishery Management Commissions, and ongoing or planned regional ocean or Great Lakes initiatives by state governors may offer opportunities for efficiently strengthening MPA collaboration, in addition to working with individual states.

The national system will use U.S. **large marine ecosystems** (LME) as the broadest framework for regional scientifically-based planning and collaboration, recognizing that certain of these regions do not efficiently or fully encompass the political regions of the United States that would be necessary for effective collaboration (Figure 5). For example, the three LMEs associated with the state and federal waters off Alaska can be combined for the purposes of regional MPA collaboration, as could the United States waters of the Caribbean and Gulf of Mexico. Nonetheless, these regions are intended to serve as the broadest framework for regional collaboration, recognizing that other established regions, whether biophysical (e.g., biogeographic regions) or political (e.g., FMC regions), may be nested within LMEs and may serve as more appropriate scales for MPA planning and collaboration. In addition, some issues, such as those pertaining to endangered and threatened species, may require regional collaboration across two or more LMEs.

The national system's regional collaboration framework will be built at the broadest level around the following regions, each encompassing state and federal waters, as relevant:

- Alaska: Gulf of Alaska, East Bering Sea, and Arctic Seas
- West Coast: California, Oregon, and Washington
- Great Lakes: Minnesota, Wisconsin, Michigan, Illinois, Indiana, Ohio, Pennsylvania, and New York
- Gulf of Mexico: Texas, Louisiana, Mississippi, Alabama, and Florida
- Caribbean: U.S. Virgin Islands, Puerto Rico, and Navassa Island
- Northeast: Virginia, Maryland, Delaware, New Jersey, New York, Connecticut, Rhode Island, Massachusetts, New Hampshire, and Maine
- Pacific Islands: Hawai'i, American Samoa, Guam, the Commonwealth of the Northern Mariana Islands, and the Pacific Remote Insular Areas (Baker Island, Howland Island, Jarvis Island, Johnston Atoll, Kingman Reef, Wake Island, and Palmyra Atoll)
- Southeast: Florida, Georgia, South Carolina, and North Carolina

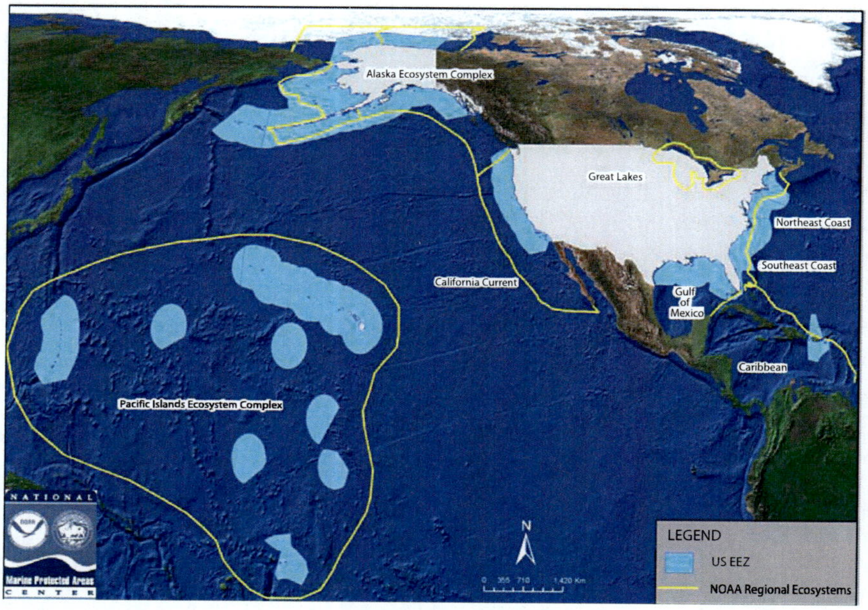

Figure 5. NOAA Regional Ecosystems of the United States

A variety of approaches exist for enhancing regional MPA coordination and collaboration. The appropriate mechanism for any particular region depends in large part on its biophysical and political characteristics and on the specific goals for which the coordination and collaboration are initiated. The MPA Center will work with all participating state, tribal, and federal MPA sites and programs and existing regional entities in each region to establish and/or formalize an appropriate regional MPA coordination and collaboration mechanism, such as a regional MPA working group, forum, or dialogue.

The MPA Center will consult with participating managing entities in the region to determine the most suitable type (e.g., informal, formal) of coordination and collaboration and the appropriate regional scale. This task includes identifying existing regional MPA and related marine coordination initiatives and determining efficient ways to complement, support or integrate with those efforts, while ensuring opportunities for all national system partners to be represented and for the public to participate.

The facilitation of formalized regional coordination and collaboration mechanisms for the national system is intended to provide a forum for MPA managing entities to work together in an open, transparent manner to:

- develop regional MPA effectiveness and stewardship strategies that identify and prioritize shared needs for improving the effectiveness of existing MPAs in the region (see Section V (A)(1));
- catalyze collaborative initiatives and projects to address identified science and stewardship needs;
- further evaluate identified national system conservation gaps, undertake collaborative, ecosystem-based MPA planning, solicit stakeholder input, and make specific recommendations about the need for the establishment of new MPAs (see Section V (A)(2));

- facilitate continued and new managerial collaboration among MPAs across regional, national, and international boundaries, to promote consistent approaches to monitoring, enforcement, emergency response, threat abatement, and coordination with other countries and international organizations (such as through transboundary MPAs) and ensure compliance with international law;
- coordinate ecosystem and/or regional input to the national system and recommend annual and longer-term regional science and other priorities based on shared MPA needs across the region;
- develop informal and formal partnerships to achieve economies of scale. For instance, arrange for the sharing of technical and financial resources for monitoring, surveillance, enforcement, staff training, etc.; and
- develop and implement strategies for engaging and informing stakeholders about regional MPA planning, effectiveness, and stewardship activities.

1. Improving MPA Stewardship, Science, and Effectiveness

A significant purpose of the Order is to "strengthen the management, protection, and conservation of existing [MPAs]..." (Section 1 (a)). As such, a major emphasis of the national system is to provide support for the shared science, technical, education, and other priority stewardship needs of partner MPA programs to enhance the national system's effectiveness. With this in mind, collaborative efforts should work to enhance the effectiveness of and provide benefits to existing efforts of MPA programs without creating additional responsibilities that detract from the important work of partners in meeting their existing programmatic authorities.

Formalizing regional coordination mechanisms via the national system offers a unique forum for collaboration to improve the effectiveness and stewardship of existing MPAs by identifying common needs across MPA programs. To this end, the MPA Center will consult with participating federal, state, and tribal managing entities through formalized regional MPA coordination and collaboration forums to develop regional MPA Stewardship, Science, and Effectiveness Strategies (Strategies). These Strategies will identify, inventory, and prioritize shared science, education, research, management, and other needs for improving MPA stewardship, science, and effectiveness. Wherever possible, these Strategies will incorporate or build upon relevant priorities previously identified through other mechanisms to avoid duplicative efforts.

The development of Strategies is intended to provide an efficient mechanism for the MPA Center to work with participating MPA sites and programs to gather information that will serve as the basis for catalyzing collaborative actions to address shared priorities. The MPA Center will also aggregate the priorities identified in the regional Strategies into a national set of priorities and use these priorities to catalyze large-scale projects and initiatives.

The following are examples of the types of priority science and stewardship issues that may be identified and addressed through the development of regional Strategies and subsequent collaborative actions among MPA programs to improve MPA effectiveness:

- Enhancing MPA management capacity
    - management plan development and review;

  - managing visitor and user impacts;
  - enforcement and compliance practices;
  - best practices for meaningful stakeholder involvement; and
  - sustainable financing mechanisms.
  -

- Improving MPA science and research
  - developing science-based tools to identify and measure regional, ecosystem, and site connectivity;
  - building collaborative strategies for establishing biophysical, social, and economic baselines for MPAs and monitoring trends in these conditions; and
  - examining the effects of invasive species on MPAs.

- Promoting outreach and education
  - developing educational programs;
  - improving awareness and understanding of the importance of marine resources and the role of MPAs in marine management; and
  - improving public stewardship of marine resources through volunteer programs and other efforts.

- Improving the evaluation of MPA effectiveness
  - training and technical assistance on developing relevant indicators and protocols for monitoring and evaluating management effectiveness for individual MPAs and networks of MPAs;
  - identifying consistent indicators for examining marine habitat and resource recovery and social and economic conditions associated with MPAs; and
  - synthesizing recovery trajectories for marine resources to aid managers, stakeholders, and the public in interpreting monitoring results and understanding habitat and resource restoration.

The Strategies will reflect shared needs, and will be implemented, subject to the availability of funds and other resources, through partnerships among MPA programs and others. Possible mechanisms to implement the Strategies could include:

- training and workshops;
- direct technical assistance and tools;
- contractual or grant funding;
- best practices or technical publications;
- sharing of knowledge and experience across MPA sites and programs;
- clearinghouse for research on MPA issues;
- targeted research;
- facilitation of linkages with international MPA programs and activities; and
- other mechanisms as identified.

2. Regional MPA Planning

The establishment or enhancement of regional MPA coordination forums via the national system offers an opportunity for managing entities and stakeholders to look beyond their individual jurisdictions, mandates, programs, and interests and consider regional and/or ecosystem-based approaches to MPA planning.

The MPA Center will work with regional, national, and international partners, where appropriate, to develop and apply the natural and social science information, decision making tools, and stakeholder engagement processes to collaboratively evaluate conservation gaps identified in the national system and make recommendations about the need for new and/or enhanced MPAs.

Such an ecosystem-based MPA planning effort could include, but is not limited to, the following critical planning steps or components:

- An evaluation and synthesis of national system design principles and conservation gaps and other regional and/or programmatic marine conservation targets, in order to more comprehensively establish regional conservation objectives to guide ecosystem-based planning.
- The characterization of marine natural resources (natural resources, habitats, ecosystems, ecological processes) and marine cultural resources in the region.
- An assessment of human uses and their impacts, including the documentation and characterization of the patterns, intensity, and significance of human uses; existing governance frameworks; and assessments of conflicts, compatibilities, and potential impacts of human uses on marine ecosystems.
- The development and use of decision tools to identify and recommend areas in need of additional or enhanced protection.
- Facilitation of stakeholder outreach and engagement processes to ensure the public and other stakeholders are informed of planning activities and have an opportunity to provide input into decision making processes.
- Development of recommendations for new or strengthened MPAs to meet regional and national priority conservation objectives and mechanisms and processes for relevant MPA authorities in establishing new MPAs or otherwise implementing recommended actions.

B. National and International Coordination

National Coordination

In addition to enhancing regional coordination among MPAs, a corresponding national level effort is needed. Such an effort will represent and promote the priorities and issues of the various ecosystems and regions that make up the nation, as well as look more broadly at important national and international trends, developments, priorities, and legal obligations. National coordination also will serve to link across regions where resource conservation issues and MPA planning and management span regional boundaries. As required by the

Order, the MPA Center will facilitate coordination at the national level. The Management Committee, described below, will be established as part of this coordination.

The Management Committee should, where possible, be composed of one representative each from a federal, state, tribal, and local government and Regional Fishery Management Council within the region, as well as the members of the Federal Interagency MPA Working Group. The committee will provide operational guidance to the national system from the perspective of MPA managers. The MPA FAC will continue to provide recommendations to DOC and DOI on the implementation of the Order and on national system implementation from a stakeholder perspective.

The Management Committee will:

- provide advice to the MPA Center on annual and long-term priorities and plans for national system support to sites and regions, based on regional stewardship and other priorities and the recommendations of the MPA FAC;
- identify management issues and other priorities that require inter-regional, national, and/or international coordination or efforts; and
- review and provide comment on conservation gaps identified at the ecosystem, regional, and/or national levels.

Regional representatives to the Management Committee will be selected by the participating MPA managing entities in the region. Each federal agency will maintain an appointed ex officio member of the Federal Interagency MPA Working Group, who also will serve on the Management Committee. Finally, two MPA FAC members, representing different stakeholder interests, will serve as ex officio members of the Management Committee.

International Coordination

In addition to U.S. MPA programs and authorities, there are numerous international MPA efforts and linkages that can contribute to and benefit from the national system. The United States shares a number of common resources with both neighboring and distant countries, and technical capabilities reside in many countries, organizations, and institutions around the world. In recognition of these important international connections, Section 4(a)(8) of the Order calls on federal agencies to identify opportunities to improve "linkages with, and technical assistance to, international [MPA] programs."

For instance, migratory species (e.g., whales, sea turtles, pelagic fishes, and birds) rely on the marine and coastal waters of multiple countries during various stages of their lives. In addition, there are also a number of international law and policy issues regarding our underwater cultural heritage. For example, certain cultural resources that rest in the seabed of U.S. MPAs, such as sunken military craft and associated contents that have not been abandoned, have a protected sovereign status and permanent right, title, and interest may be vested in the flag country.

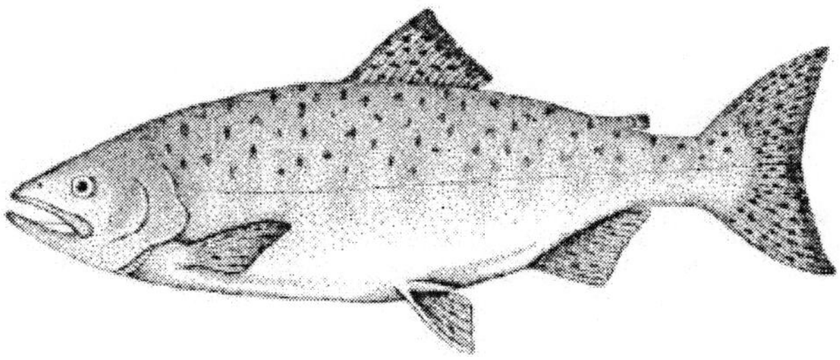

To strengthen international coordination on MPA issues, the MPA Center, representing the National System of MPAs, and the Management Committee, in coordination with the U.S. Department of State and internationally relevant regional forums, can seek to enhance existing or establish new linkages with efforts in other countries, in accordance with international law. Such linkages should be focused on issues of mutual benefit to U.S. and international MPAs and MPA programs, such as policy coordination, collaborative activities, information and capacity sharing, capacity building, and technical assistance

C. Evaluating National System Effectiveness

Monitoring and evaluating management effectiveness is a key component of an effective, adaptively managed national system. To this end, the Order calls for "practical, science-based criteria and protocols for monitoring and evaluating the effectiveness of MPAs" (Section 4(a)(5)). Monitoring and evaluation efforts of the national system are focused on measuring the effectiveness of the national system in achieving its priority conservation objectives and management objectives and the contributions of participating national system MPAs and MPA programs in achieving those objectives. It is not a function of the national system to monitor or evaluate individual MPAs or MPA programs, although the national system can provide assistance to MPA programs to assist them in better evaluating their own efforts. Stakeholders with an interest in participating in the monitoring of individual MPAs or MPA programs should consult with the managing entity or entities.

The national system's approach to evaluating effectiveness will include:

- technical and scientific support for fostering sound monitoring, and evaluation programs at the participating MPA site or system level;
- development and implementation of a set of standards and protocols for assessing broader national system effectiveness. In order to be efficient and effective, the development of such standards and protocols requires significant input and advice from participating national system MPA sites and systems; and
- cooperation with existing or developing observation, monitoring and evaluation programs.

The natural and social science data currently collected and used by MPA sites and systems to monitor and evaluate their own effectiveness will not only help in their adaptive management efforts, but also will contribute to the analysis of the national system's success in meeting its goals. The national system will aim to support the tools and technical assistance needed by partner MPA sites and systems to effectively monitor and evaluate their own effectiveness. It will not create new requirements for sites or systems to undertake new or expanded monitoring and evaluation activities.

With advice from the MPA FAC, the Management Committee, national system MPA partners in the regions, and science and management experts, the MPA Center will develop and publish guidance for monitoring and evaluating the effectiveness of the national system. These guidelines will provide an integrated approach for monitoring the effectiveness of the national system, including the degree to which the priority conservation objectives are met and the benefits are provided to participating MPA sites and systems.

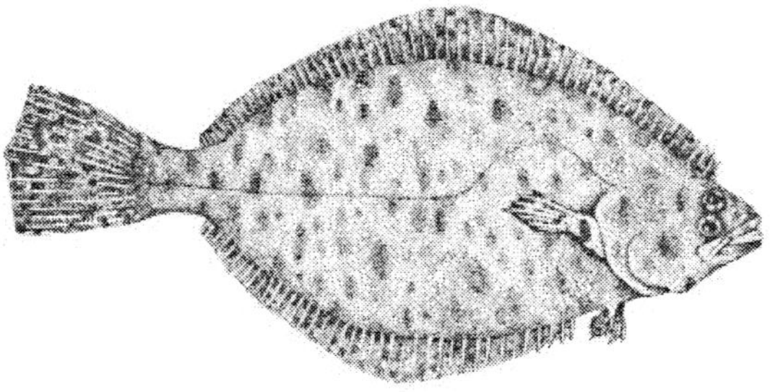

In addition, if identified as stewardship priorities by participating MPA sites and systems, training and technical assistance efforts targeted at monitoring and evaluation can be developed, such as establishing relevant sets of natural and social science indicators and protocols.

The results of monitoring and evaluating the national system will be used to manage the system adaptively and identify future focus areas for stewardship and other initiatives, including but not limited to: conservation gaps; technical and other forms of assistance in support of MPA sites and programs; and necessary changes to the national system's goals, objectives, or other components.

D. Federal Agency Responsibilities to Avoid Harm

Section 5 of the Order calls for federal agencies to "avoid harm" to the natural and cultural resources protected by MPAs that become part of the national system. Each federal agency is responsible for its own implementation of its responsibilities under Section 5.

The Order states:

Each Federal agency whose actions affect the natural or cultural resources that are protected by an MPA shall identify such actions. To the extent permitted by law and to the maximum extent practicable, each Federal agency, in taking such actions, shall avoid harm to the natural and cultural resources that are protected by an MPA. In implementing this section, each Federal agency shall refer to the MPAs identified under subsection 4(d) of this order.

Implementation

To implement Section 5 of the Order:

- The MPA Center will collect, maintain, and make publicly available via the MPA Center's website, http://www.mpa.gov, and *Federal Register* notices, all relevant regulatory and resource information for MPAs that are subject to agency requirements under Section 5, in the form of a List of National System MPAs. National system MPAs included in the List are those that have satisfied the requirements outlined in Sections III (B) and (D) of the Framework and are officially a part of the National System of MPAs. Information maintained for each national system MPA on the List will include: site name, location, national system MPA category, priority conservation objective(s) contributed to, boundaries, key resources protected, authorizing legislation, level and types of protection, managing authority/program, name of point of contact, and relevant contact information.
- Federal agencies shall: (1) identify their activities that affect the natural or cultural resources protected by individual national system MPAs, and (2) to the extent permitted by law and to the maximum extent practicable, avoid harm to those resources. Both of these activities should be accomplished through existing natural or cultural resource management or review authorities and procedures, including, but not limited to those under:

    - National Environmental Policy Act;
    - Coastal Zone Management Act;
    - National Historic Preservation Act;
    - Endangered Species Act;
    - Federal Water Pollution Control Act (Clean oo Water Act);
    - Marine Mammal Protection Act;
    - National Wildlife Refuge System Administration Act;
    - National Park Service Organic Act;
    - Rivers and Harbors Act;
    - Sunken Military Craft Act;
    - National Marine Sanctuaries Act (Title III oo of the Marine Protection, Research, and Sanctuaries Act);
    - Magnuson-Stevens Fishery Conservation oo and Management Act;
    - Outer Continental Shelf Lands Act;
    - Coral Reef Conservation Act;
    - Energy Policy Act of 2005; and
    - Other pertinent statutes and Presidential Executive Orders.

- Upon receipt of a federal agency's request for assistance, the MPA Center will work to facilitate support for policy and coordination assistance through existing agency review processes.
- As needed, the MPA Center, working with federal agencies, will produce voluntary technical guidance and best practices on priority issues to assist federal agencies in

their determination of impacts to marine resources protected by national system MPAs and options for avoiding harm. The MPA Center also will work with federal agencies to provide clear public outreach materials to educate and inform the public on the requirements of Section 5.

- Federal agencies will report their actions to implement Section 5, any comments received, and responses to such comments on an annual basis as part of the agency report required by Section 6 of the Order. The MPA Center, as required by the Order, will post these reports on the http://www.mpa.gov website.

Activities to Be Considered

The implementation of Section 5 is governed by existing authorities, each with its own threshold and/or trigger for requiring individual federal agencies to identify, review, mitigate, or otherwise alter their activities based on impacts to natural or cultural resources. The Order does not provide any new authority for any federal agency or the MPA Center to review activities of any other federal agency or alter standards for existing review. The thresholds and/or triggers for agency action under Section 5 are the same as those listed under any existing authority or authorities that normally require agency review of a proposed activity. Section 5 does, however, require agencies to ensure that their activities avoid harm to the natural and cultural resources as protected by the MPAs included in the national system (to the extent permitted by law and to the maximum extent practicable) when fulfilling their existing requirements for identifying, reviewing and implementing activities.

Furthermore, there is no single definition for key terms used to describe the requirements under Section 5, including but not limited to: "avoid harm," "affect," or "to the extent permitted by law and to the maximum extent practicable." Instead, the meaning of any of these terms, as applied to an agency's requirements under Section 5, is dependent on the agency's interpretation, consistent with any requirements of the legal framework used to protect the resources of the MPA and any other applicable natural or cultural resource review or protection authorities or procedures.

Pursuant to Section 5 of the Order, agency requirements apply only to the natural or cultural resources specifically afforded protection by the site as described on the List of National System MPAs. For example, within national system MPAs established for sustainable production, other resources not specifically protected by the MPA would not be subject to the "avoid harm" provision. For sites that have both a terrestrial (i.e., an area that falls outside of the definitional boundaries of 'marine') and marine area, only the marine portion and its associated protected resources will be included on the List of National System MPAs and subject to Section 5 of the Order. To implement Section 5, each federal agency shall identify its activities that affect the natural or cultural resources protected by a national system MPA through the existing natural and cultural resource review processes normally required for these activities. Similarly, the determination of whether an agency in taking such actions is avoiding harm to those resources, to the extent permitted by law and to the maximum extent practicable, will be made by the individual agency using its existing natural and cultural resource review processes and/or authorities.

Comment and Response on Agency Actions

Comments from any person, organization, or government entity concerning federal agency compliance with Section 5 should be directed to the relevant lead federal agency for the action or actions that are the subject of the comments. Each agency shall make a determination on the response and take appropriate action. Similarly, any requests for information regarding compliance with Section 5, including those under the Freedom of Information Act (FOIA), should be directed to the lead agency for the action or actions that are the subject of the request. Any comments or requests for information received by the MPA Center or any federal agency in regard to another agency's compliance with this Section shall, pursuant to FOIA procedures, be forwarded in a timely manner to the relevant responsible agency for its consideration, with due notice given to the sender.

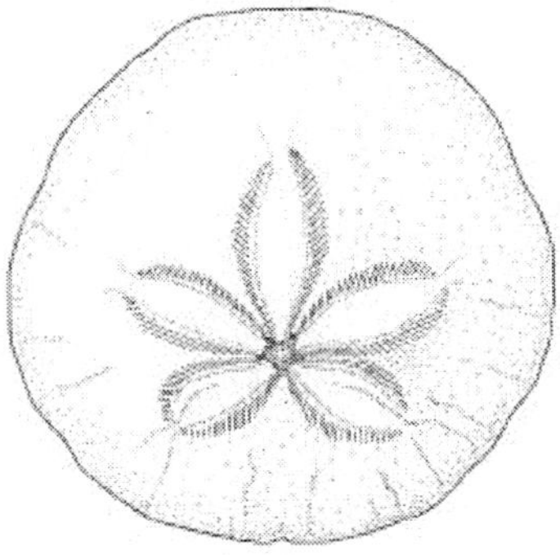

Reporting and Periodic Review

As required under Section 6. Accountability of the Order, "[e]ach Federal agency that is required to take actions under the order shall prepare and make public annually a concise description of actions taken by it in the previous year to implement the order, including a description of written comments by any person or organization stating that the agency has not complied with this order and a response to comments by the agency." These annual reports, including a point of contact for each federal agency, will be posted at http://www.mpa.gov. In addition, on a biennial basis, the MPA Center will consolidate agency annual reports into a biennial "State of the National System of MPAs" report. The biennial report will include an assessment of overall progress to develop the National System of MPAs and the effectiveness of meeting its stated goals and objectives, including those related to Section 5 of the Order. More information on the biennial report can be found below in Section V (E) of this document.

E. Tracking and Reporting

Tracking and reporting of the national system are important activities for communicating regional and national accomplishments and priority future efforts in need of support. In order to track and report progress, the MPA Center will coordinate a biennial "State of the National System of MPAs" progress report and post all available data and assessments on the http://www.mpa.gov website. In addition, the MPA Center will work with the Management Committee and participating MPA sites and programs to determine how best to comprehensively track overall national system priorities once efforts to establish the sytem have been initiated. Additional information on these efforts is provided below.

Biennial "State of the National System of MPAs" Progress Report

On a biennial basis, the MPA Center, working with its national system partners, will develop and publish on the http://www.mpa.gov website a consolidated "State of the National System" progress report, in accordance with Section 6 of the Order. The report will consolidate and summarize the annual reports submitted by federal agencies for the period and also will include information from states and other management entities. It will include:

- a list of existing National System MPAs and newly added or removed sites;
- a summary of federal activities taken in support of the national system;
- a summary of regional, national, and international planning efforts;
- a summary of assistance provided to national system MPAs;
- an evaluation of the effectiveness of the national system in meeting its goals and objectives at the national and regional levels;
- a summary of actions taken to implement Section 5 of the Order;
- any recommendations developed by the MPA FAC during the period;
- a description of public comments received and responses sent during the period; and
- regional, national, and international priorities for future coordination, planning, technical, and other types of support (see Sections V (A) and (B) of this document).

MPA.gov Website

As required by the Order, the website http://www.mpa.gov will be maintained to communicate and archive all information about the development and implementation of the national system. The website will house information about a variety of technical, scientific, governance, and other MPA topics relevant to the breadth of MPA stakeholders, including the MPA FAC. In addition, the website will house information on national system progress, priorities, and plans, including:

- MPAs found to be eligible for nomination to the national system;
- MPAs and MPA systems that have been included in the national system;
- areas and resources identified as national system conservation gaps;
- recommendations for new or enhanced MPAs resulting from regional MPA planning;
- regional MPA science, stewardship, and effectiveness strategies and national and other priorities for improving stewardship and effectiveness;
- international activities and commitments;
- information on the nomination process and supporting analyses;
- information related to the evaluation of national system effectiveness;
- agency and MPA Center reports;

- public comments received on MPA □□nominations to and removals from the national system; and
- the official List of National System MPAs.

F. MPA Federal Advisory Committee

The MPA FAC is authorized by the Order to provide expert advice and recommendations to DOC and DOI on the development and implementation of the National System of MPAs and implementation of the Order. The MPA FAC is comprised of 30 non-federal members representing regionally diverse perspectives and areas of expertise from all regions of the country, including natural and social science, commercial and recreational fishing, tribal and state governments, oil and gas, tourism, environmental organizations, and others. It also includes ex officio members from pertinent federal agencies. A full description of the MPA FAC can be found in Appendix B and a list of the MPA FAC members, past and present, can be found in Appendix E of this document.

Throughout the development and implementation of the national system, the MPA FAC will continue to advise DOC and DOI on priority topics and issues as identified by the agencies. The MPA FAC also will provide recommendations to the MPA Center concerning national system conservation gaps, as described in Section IV (D) above.

G. Role of the National MPA Center in the National System

The specific roles of the MPA Center in coordinating the national system are to:

- provide coordination and facilitation of the national system as a whole (individual MPA programs and managing entities remain responsible for administering their sites and systems);
- coordinate processes to identify, nominate, and include eligible MPAs in the national system, remove MPAs from the national system, and maintain the List of National System MPAs;
- build public and private partnerships and catalyze action to support the identified science, stewardship, and effectiveness priorities of participating MPA programs;
- facilitate the development and maintenance of regionally appropriate MPA coordination mechanisms among participating programs, and, where possible, maintain a Regional MPA Coordinator in the field to support such efforts;
- develop, in consultation with participating programs, regional MPA Science, Stewardship, and Effectiveness Strategies;
- lead collaborative efforts to identify conservation gaps in the national system;
- build and catalyze partnerships and actions to provide technical or scientific information, staff, or other support for collaborative ecosystem-based MPA planning in order to identify and recommend new or enhanced MPAs;
- promote stewardship of the national system through effective outreach and education;

- support the operation of the MPA FAC and the coordination of the MPA Federal Interagency Working Group and Management Committee;
- track, communicate, integrate, and recommend suggested MPA science and other national system priorities, needs, and commitments across the regional, national, and international levels;
- develop a biennial "State of the National System of MPAs" report and maintain comprehensive information about the national system's priorities and progress on the http://www.mpa.gov website;
- monitor and evaluate the effectiveness of the national system and implement adaptive management strategies based on results; and
- maintain the http://www.mpa.gov website as a mechanism for communicating information about the national system.

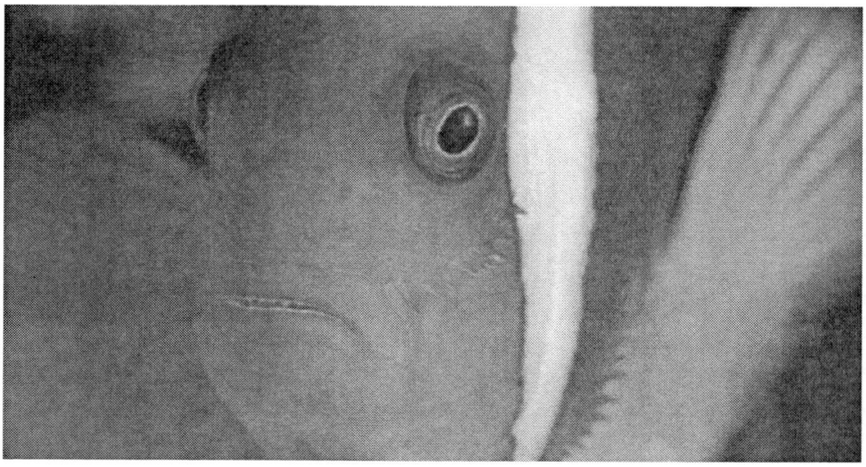

VI. Glossary of Key Terms

The following are definitions of key terms as used in this Framework document. See Table 2 for the full definition of key terms used in the definition of an MPA.

Adaptive management – "A systematic process for continually improving management policies and practices by learning from the outcomes of operational programs." (British Columbia Forest Service, http://www.for.gov.bc.ca/hfp/amhome/Amdefs.htm).

Area – Must have legally defined geographical boundaries and may be of any size, except that the site must be a subset of the United States federal, state, local, or tribal marine environment in which it is located.

Biodiversity – The variety of living organisms in all their forms. Technically, biodiversity includes variety at three levels of biological organization: genetic variation within species, the variety of species, and the variety of ecological communities.

Conservation area – Multiple uses allowed; however, uses and activities may be restricted or zoned and access limited, as necessary to meet site management goals.

Cultural heritage – The cultural resources that reflect the nation's maritime history and traditional cultural connections to the sea, and the uses and values they provide to present and future generations.

[Marine] Cultural resource – A tangible entity that is valued by or significantly representative of a culture, or that contains significant information about a culture. Cultural resources for purposes of the MPA Executive Order are tangible entities at least 50 years in age that reflect the nation's maritime history and traditional cultural connections to the sea, such as archaeological sites, historic structures, shipwrecks, artifacts, and traditional cultural properties. Cultural resources are categorized as districts, sites, buildings, structures, and objects for the National Register of Historic Places, and as archaeological resources, cultural landscapes, structures, and ethnographic resources for MPA management purposes. Ethnographic resources include natural resources and sites with tribal or traditional cultural meaning, value and use.

Ecological network – A set of discrete MPAs within a region that are connected through dispersal of reproductive stages (eggs, larvae, spores, etc.) or movement of juveniles and adults. The effective management of certain marine species may require networks of discrete MPAs encompassing regional collections of local populations linked by dispersal and movement, which may be essential for some local populations to persist. The creation of MPA networks must take into consideration other non-MPA areas that provide similar linkages, which does not necessarily imply additional management measures outside MPAs or the creation of a "super MPA" with boundaries encompassing all MPAs in the network.[9]

Ecosystem – A geographically specified system of organisms, including humans and the environment and the processes that control its dynamics.

Ecosystem approaches to management (or Ecosystem-based management) – A management approach that "looks at all the links among living and nonliving resources, rather than considering single species in isolation." This approach "reflects the relationships among all ecosystem components, including humans and nonhuman species, and the environments in which they live. This system of management considers human activities, their benefits, and their potential impacts within the context of the broader biological and physical environment."[10]

Extractive – Activities that remove or are intended to remove living or nonliving resources from an MPA.

Large Marine Ecosystems – Regions of ocean space encompassing coastal areas from river basins and estuaries out to the seaward boundary and continental shelves and the seaward margins of coastal current systems. They are relatively large regions on the order of 200,000 square kilometers or greater, characterized by distinct bathymetry, hydrography, productivity, and trophically dependent populations.

Lasting – For natural heritage and cultural heritage MPAs, the site's authority must clearly state its intent to provide permanent protection. For sustainable production MPAs, the site must be established with the intent at the time of designation to provide, at a minimum, the duration of protection necessary to achieve the mandated long-term sustainable production objectives for which the site was established.

Local government – A legally established unit of government at a level below state government, including but not limited to county, city, town, or village.

Management [managing] entity or entities – The federal, state, local, or tribal entity or entities with legal authority to designate, promulgate regulations for, and/or manage an MPA. In many cases, authority lies with one entity or program; however, in certain instances, such as the federal/state National Estuarine Research Reserve System and state/tribe co-management arrangements, authority is formally shared or split among two or more entities.

Marine environment – Must be: (a) ocean or coastal waters (note: coastal waters may include intertidal areas, bays, or estuaries); (b) an area of the Great Lakes or their connecting waters; (c) an area of lands under ocean or coastal waters or the Great Lakes or their connecting waters; or (d) a combination of the above.

Marine Protected Area – Any area of the marine environment that has been reserved by federal, state, territorial, tribal, or local laws or regulations to provide lasting protection for part or all of the natural and cultural resources therein. See also Area, Marine environment, Reserved, Lasting, and Protection.

Marine Reserve – A type of MPA where extractive uses are prohibited (also referred to as "no-take" reserve).

National System of MPAs – The group of MPA sites, networks, and systems established and managed by federal, state, tribal, and/or local governments that collectively enhance conservation of the nation's natural and cultural marine heritage and represent its diverse ecosystems and resources. National system MPAs work together at the regional and national levels to achieve common objectives for conserving the nation's important natural and cultural resources.

Natural heritage – The nation's biological communities, habitats, ecosystems, and processes and the ecological services, uses, and values they provide to present and future generations.

[Marine] Natural resource – Any biological or physical component of the marine environment that contributes to the structure, function, goods, or services provided by a marine ecosystem.

Network – A set of discrete MPAs within a region or ecosystem that are connected through complementary purposes and synergistic protections. A network of MPAs could focus on ecosystem processes, certain individual marine species, or cultural resources. For example, an ecological network of MPAs could be connected through dispersal of reproductive stages or movement of juveniles and adults (see "Ecological network").

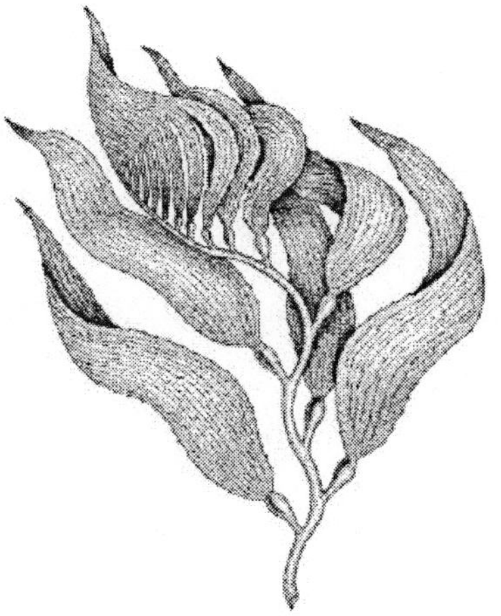

Precautionary design – Decisions are based on the best information currently available from natural science, social science, customary and local knowledge, and other sources. Where information is limited, decisions should reflect a precautionary approach.

Protection – Must have existing laws or regulations that are designed and applied to afford the site with increased protection for part or all of the natural and submerged cultural resources therein for the purpose of maintaining or enhancing the long-term conservation of these resources, beyond any general protections that apply outside the site.

Region or Regional – An area inclusive of and determined by participating national system sites and systems that is based on common management interests, similar or linked ecological characteristics, and/or other factors that provide a foundation for meaningful coordination.

Reserve area – No extractive uses allowed, except permitted scientific and educational uses; destructive or disruptive activities are limited; other uses and activities may be restricted or zoned; and access is limited, as necessary to meet site management goals.

Reserved – Must be established by and currently subject to federal, state, local, or tribal law or regulation.

Stakeholder – Individuals, groups of individuals, organizations, or political entities interested in and/or affected by the outcome of management decisions. Stakeholders may also be individuals, groups, or other entities that are likely to have an effect on the outcome of management decisions. Members of the public also may be considered stakeholders.

State – See United States.

Stewardship – Careful and responsible management to ensure goals and objectives are being achieved for the benefit of current and future generations.

Sustainable production resources – The nation's renewable living resources and their habitats (including, but not limited to, spawning, mating, and nursery grounds and areas established to minimize bycatch of species) and the social, cultural, and economic values and services they provide to present and future generations.

System – A set of MPAs connected by shared programmatic, administrative, or other organizing principles or purposes. A system of MPAs is not necessarily confined to a specific geographic area such as a region or ecosystem.

Tribe – A federally recognized American Indian or Alaska Native government.

United States – Includes the several states, the District of Columbia, the Commonwealth of Puerto Rico, the Virgin Islands of the United States, American Samoa, the Commonwealth of the Northern Mariana Islands, and Guam.

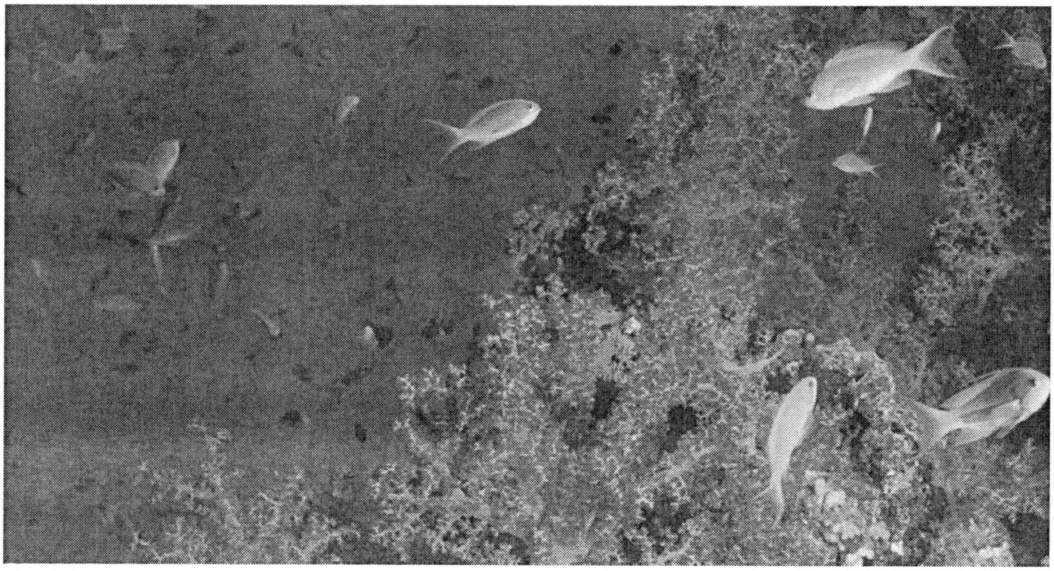

VII. Administrative and National Policy Requirements

- Nothing in Executive Order 13158 or this Framework shall be construed as altering existing authorities regarding the establishment of federal MPAs in areas of the marine environment subject to the jurisdiction and control of states, the District of Columbia, Commonwealth of Puerto Rico, the U.S. Virgin Islands, American Samoa, Guam, and the Commonwealth of the Northern Mariana Islands.
- Neither Executive Order 13158 nor this Framework creates any right or benefit, substantive or procedural, enforceable in law or equity by a party against the United States, its agencies, its officers, or any person.
- Neither Executive Order 13158 nor this Framework diminishes, affects, or abrogates Indian treaty rights or U.S. trust responsibility to Indian tribes.

- Federal agencies taking actions pursuant to Executive Order 13158 or under this Framework must act in accordance with international law and with Presidential Proclamation 5928 of December 27, 1988, on the Territorial Sea of the United States of America; Presidential Proclamation 5030 of March 10, 1983, on the Exclusive Economic Zone of the United States of America; and Presidential Proclamation 7219 of September 2, 1999, on the Contiguous Zone of the United States.

APPENDIX A. ACRONYMS AND ABBREVIATIONS USED

Acronyms

COP	–	Commission on Ocean Policy
DOC	–	Department of Commerce
DOI	–	Department of the Interior
EPA	–	Environmental Protection Agency
FOIA	–	Freedom of Information Act
FMC	–	Federal Fishery Management Council
FWS	–	U.S. Fish and Wildlife Service
LME	–	Large Marine Ecosystem
MLCD	–	Manele-Hulopoe Marine Life Conservation District
MPA	–	Marine protected area
MPA FAC	–	Marine Protected Areas Federal Advisory Committee
NEPA	–	National Environmental Policy Act
NRCE	–	National Register Criteria for Evaluation
NRHP	–	National Register of Historic Places
NERRS	–	National Estuarine Research Reserve System
NMFS	–	National Marine Fisheries Service
NOAA	–	National Oceanic and Atmospheric Administration
NRCE	–	National Register Criteria for Evaluation
NRHP	–	National Register of Historic Places
SIMOR	–	Subcommittee on Integrated Management of Ocean Resources

U.S.	–	United States of America
USOAP	–	U.S. Ocean Action Plan (USOAP)
USGS	–	US Geological Survey
WCPA/IUCN	–	World Commission on Protected Areas/International Union for Conservation of Nature

Abbreviations

Framework	–	Framework for Developing the National System of MPAs
MPA Center	–	National Marine Protected Areas Center
National System	–	National System of Marine Protected Areas
NOAA Fisheries Service	–	NOAA's National Marine Fisheries Service
Order	–	Executive Order 13158 of May 26, 2000
Management Committee	–	National System Management Committee
Strategy	–	MPA Stewardship, Science and Effectiveness Strategy

Appendix B. Existing U.S. MPA Programs, Federal MPA Initiatives, and Tribal and International Efforts

The nation's existing suite of MPA sites, programs, authorities, and systems at all levels of government are the fundamental components of the national system. The recognition of and full participation by these federal, state, tribal, and local government programs are critical to the national system's success. Working together, these existing programs and authorities, federal MPA coordination initiatives, and linkages to international MPA initiatives will make important contributions to and receive benefits from the development of an effective national system. This section provides an overview of these major efforts and generally describes their respective roles in the national system.

A. U.S. MPA Programs and Authorities

MPAs in the United States are managed by a number of entities and programs at federal, state, tribal, and local government levels. This section provides a brief summary of these programs and describes the nature of their role in the development of the national system.

Federal and Federal/State MPA Programs

Currently, there are several federal MPA programs and one federal/state partnership MPA program in the United States. Each has one or more specific legal mandates that it is required to fulfill. Many of these programs have established and actively manage systems of MPAs designed to fulfill their responsibilities to the nation. As described below, the federal

MPA programs include DOI's National Park System and National Wildlife Refuge System and NOAA's National Marine Sanctuary System, National MPA Center, and National Marine Fisheries Service programs. The National Estuarine Research Reserve System is composed of NOAA/state partnerships.

National Park System: The National Park System is administered by DOI's National Park Service with a mission to conserve the scenery and the natural and historic objects and wildlife therein and to provide for the enjoyment of the same in such manner and by such means as will leave them unimpaired for the enjoyment of future generations. The National Park System preserves unimpaired natural and cultural resources and values representative of the nation's ocean heritage in superlative natural, historic, and recreation areas in every region. The National Park System currently contains 72 ocean and Great Lakes parks.

National Wildlife Refuge System: The U.S. Fish and Wildlife Service's (FWS) mandate is to provide the federal leadership to conserve, protect, and enhance fish and wildlife and their habitats for the continuing benefit of people. The mission of the National Wildlife Refuge System, a program within the DOI FWS, is to administer a national network of lands and waters for the conservation, management, and where appropriate, restoration of the fish, wildlife, and plant resources and their habitats within the United States for the benefit of present and future generations of Americans. There are 177 ocean and Great Lakes refuges.

National Marine Sanctuary System: Under the National Marine Sanctuaries Act, NOAA establishes areas of the marine environment that have special conservation, recreational, ecological, historical, cultural, archaeological, scientific, educational, or aesthetic qualities as national marine sanctuaries to: (A) improve the conservation, understanding, management, and wise and sustainable use of marine resources; (B) enhance public awareness, understanding, and appreciation of the marine environment; and (C) maintain for future generations the habitat and ecological services of the natural assemblage of living resources that inhabit these areas. There are currently 13 sanctuaries and one marine national monument in the national marine sanctuaries system.

National Marine Protected Areas Center (MPA Center): The mission of the MPA Center is to facilitate the effective use of science, technology, training, and information in the planning, management, and evaluation of the nation's system of marine protected areas. The MPA Center is housed within NOAA and coordinates across NOAA programs, as well as with pertinent federal, state, tribal, and local MPA and MPA-support entities. At the federal level, the MPA Center coordinates closely with DOI. The MPA Center's specific national system roles are described in detail in Section V (G) of this document.

National Marine Fisheries Service Programs and Federal Fishery Management Councils (FMC): Under a number of statutory authorities, the National Marine Fisheries Service establishes and manages MPAs to rebuild and maintain sustainable fisheries, conserve and restore healthy marine habitats, and promote the recovery of protected species, including marine mammals and anadromous fish. These sites fall under four major categories: Federal Fisheries Management Zones, Federal Fisheries Habitat Conservation Zones, Federal Threatened and Endangered Species Protected Areas, and Federal Marine Mammal Protected

Areas. FMCs have been established for the stewardship of fishery resources through the preparation, monitoring, and revision of fishery management plans. These FMCs enable states, the fishing industry, consumer and environmental organizations, and other interested persons to participate in and advise on the management of marine fisheries and to take into account the social and economic needs of the states. FMC-recommended actions are subject to review and approval by the Secretary of Commerce through a delegation of authority to the National Marine Fisheries Service. The National Marine Fisheries Service is responsible for the promulgation of site-specific regulations to delineate MPA boundaries and establish associated protective measures.

National Estuarine Research Reserve System (NERRS): The mission of the NERRS is to promote stewardship of the nation's estuaries through science and education using a system of protected areas. The NERRS, which is currently made up of 27 sites, is a unique partnership program between NOAA and the coastal states to protect estuarine land and water, which provides essential habitat for wildlife, and offers educational opportunities for students, teachers, and the public. The NERRS sites serve as living laboratories for scientists. With its unique state/federal partnership, the NERRS participation with the national system will require close consultation and coordination with the NOAA Estuarine Reserves Division and state agency or university staff of NERRS sites.

National Monuments: In June 2006, President Bush established the Northwestern Hawaiian Islands Marine National Monument under Presidential Proclamation 8031 (71 FR 36443, June 26, 2006) under the authority of the Antiquities Act (16 U.S.C. 431). This was the nation's first marine national monument. The Monument–renamed the Papahānaumokuākea Marine National Monument in March 2007 to reflect Hawaiian language and culture – is approximately 100 nautical miles wide and extends approximately 1,200 miles from northwest to southeast around the Northwestern Hawaiian Islands. In December 2006, the Secretaries of Commerce and the Interior and the Governor of Hawai'i signed a Memorandum of Agreement to jointly manage federal and state lands and waters within the Monument as Co-Trustees, to collectively conserve and manage Monument natural and cultural resources.

State and Local Government MPA Programs

Each U.S. coastal state also has a variety of MPA programs and authorities, often at both the state and local government levels. State MPA programs can include: Historic Preservation offices; Fish and Wildlife agencies; Coastal Zone Management programs; Fishery Management agencies; Parks and Recreation agencies, and other authorities. MPAs are used by states for a variety of purposes ranging from managing fisheries, recreation, tourism, and other uses to protecting ecological functions, preserving shipwrecks, and maintaining traditional or cultural connections to the marine environment. In addition, local governments within coastal states, such as counties and other municipalities, have programs that establish and manage MPAs for protecting marine species, nursery grounds, shellfish beds, and other important natural and cultural resources. Similar to their federal analogs, some state MPA programs have also developed and continue to manage their existing sites as systems of MPAs.

Given the significant coastal and marine resources under state jurisdiction, the large number of state MPAs – roughly 83 percent of the national total – compared to federal sites, and the potential impacts and benefits to states from MPAs located in federal waters, full state participation in the development of the national system is critical to its success. It is important to note, however, that state and local government participation in the national system is voluntary under the Order. The MPA Center will work closely with states to determine their interest in participating. State government agencies, programs, and authorities that elect to participate in the national system will be full partners and will have an equal voice in decision making to set priorities for collaborative efforts at the regional and national levels.

Tribal MPA Authorities, Programs, and Linkages

Tribal governments have an integral role to play in resource management—legally, culturally and economically. The Order "does not diminish, affect, or abrogate Indian treaty rights or United States trust responsibilities to Indian tribes," and calls on NOAA and DOI to "consult with…tribes…and other entities to promote coordination of federal, state, territorial, and tribal actions to establish and manage MPAs." Because the federal government has a trust responsibility to all federally recognized tribes, conservation goals and management practices for MPAs should be established through government-to-government consultations.

In addition, several Indian tribes in Western Washington and the Great Lakes have treaty-reserved fishing rights. These tribes share co-management authority and responsibility for marine resources in their usual and customary fishing areas with the federal government and/or states, depending on the specific resource and area identified. Tribes that have sole management authority may choose to establish MPAs as a tool to meet conservation goals for areas where they have management responsibilities. For areas where tribes share co-management authority with the federal government and/or states, any entity wishing to establish MPAs must do so through government-to-government consultations. The MPA Center will work closely with tribes to determine their interest in participating in the national system. Tribal governments that elect to participate in the national system will be full partners and will have an equal voice in decision making to set priorities for collaborative efforts at the regional and national levels.

Numerous opportunities to enhance coordination and collaboration with tribes on issues related to MPAs are possible through the development of the national system. Some of these opportunities could include a range of potential partnerships aimed at the sharing of information; enhancing technical, scientific, and management capacity; and developing conservation strategies for marine resources of mutual concern. The MPA Center and national system partners, many of whom have ongoing relationships with tribes, will consult with tribal governments to determine their interest in participating in the national system and will work with them to develop appropriate mechanisms and protocols.

B. Linkages to Related Federal MPA Initiatives

There are several other significant federal MPA initiatives that are either directly or indirectly linked to the development of the national system. These efforts make important contributions to and can benefit from the development of the national system. This section provides an overview of each of these efforts and further describes their relationship and role in the development of the national system.

MPA Federal Advisory Committee

The MPA FAC is authorized by the Order to provide expert advice and recommendations to DOC and DOI. The MPA FAC is comprised of 30 non-federal members representing diverse perspectives and areas of expertise, including natural and social science, commercial and recreational fishing, tribal and state governments, oil and gas, tourism, environmental organizations, and others. The MPA FAC also includes ten federal ex officio members to provide information and support from entities managing, supporting, or potentially affecting MPAs. The MPA FAC completed its first report in June 2005, which provided recommendations on the goals, objectives, principles, and structure of the national system, and its second report in October 2007, which provided recommendations regarding the development of the national system. The MPA FAC will continue to advise DOC and DOI on aspects of developing and implementing the national system. Information on MPA FAC members and its work products are posted at http://mpa.gov/mpafac/fac.html.

The Federal Interagency MPA Working Group

The Order directs DOC and DOI to work closely with the other federal agencies to develop the national system. To provide a mechanism for this coordination, the MPA Center established the Federal Interagency MPA Working Group, which includes representatives from the Departments of Commerce, the Interior, Defense, Homeland Security, State, Agriculture, Environmental Protection Agency, National Science Foundation, and the U.S. Agency for International Development. The Federal Interagency MPA Working Group meets several times a year to provide input on policy issues related to national system development, coordinate activities related to the Order, and support the work of the MPA FAC. In addition, members of the Federal Interagency MPA Working Group will serve as members of the National System Management Committee (see Section V (B)).

U.S. Ocean Action Plan

The U.S. Ocean Action Plan (USOAP) outlines a variety of actions for promoting the responsible use and stewardship of ocean and coastal resources for the benefit of all Americans. A Cabinet-level "Committee on Ocean Policy" (COP) was established by Executive Order 13366 (December 17, 2004) to coordinate the activities of executive branch departments and agencies regarding ocean-related matters in an integrated and effective manner to advance the environmental and economic interests of present and future generations of Americans. The President further directs the executive branch agencies to facilitate, as appropriate, coordination and consultation regarding ocean-related matters among federal, state, tribal, and local governments; the private sector; foreign governments; and international organizations. Subcommittees of the COP also have been formed as part of the ocean governance structure described in the USOAP, including the Subcommittee on

Integrated Management of Ocean Resources (SIMOR) and the Joint Subcommittee on Ocean Science and Technology. Many of the activities outlined in the USOAP and the subsequent work plans of the COP's subcommittees complement efforts to develop the national system. Similarly, many of the collaborative actions under the national system may offer opportunities to help advance the USOAP. As these efforts proceed, the MPA Center will work closely with SIMOR to evaluate progress and plans for developing the national system in order to ensure coordination and consistency with the USOAP's governance structure and overall approach.

In support of this effort, the USOAP calls on National Parks, National Wildlife Refuges, National Marine Sanctuaries, and National Estuarine Research Reserves to, "coordinate and better integrate the existing network of marine managed areas." Many of these sites overlap or lie adjacent to each other and a history of collaboration between parks, marine sanctuaries, refuges, and reserves provides a model for this expanded network. Although these sites were created under separate agency authorities and statutory mandates, they are united by their proximity and similar science and management priorities. These actions to coordinate and better integrate efforts have been aptly named and are referred to as the "Seamless Network" initiative. The Seamless Network concept reflects the Administration's emphasis on greater scientific and programmatic coordination between ocean agencies, and complements efforts to implement the MPA Executive Order. In addition, the USOAP calls on the National Park Service to adopt an Ocean Parks Stewardship Action Plan. Both the Seamless Network and Ocean Parks Stewardship Action Plan are described below.

Seamless Network Initiative

The USOAP calls on the four above mentioned MPA systems to work together, "to promote coordination of research, public education, and management activities at neighboring parks, refuges, sanctuaries, and estuarine reserves." Two federal interagency agreements are called for under this effort. The first is a general agreement that enables site-based, regional, and national collaborations among the partner agencies, and is currently under development. The second is a separate cooperative enforcement agreement signed in August 2005 among the National Wildlife Refuge System, National Park Service, National Marine Sanctuary Program, and National Marine Fisheries Service. When implemented, these agreements will ultimately contribute to several important elements of the national system, such as the identification of science and stewardship priorities for enhancing MPA effectiveness through enhanced interagency cooperation and information sharing. Known as the Seamless Network initiative, this effort will provide a coordination mechanism for these MPA systems in the development of the national system and will build on existing collaborative efforts. In many cases these MPAs have ongoing collaborations and the Seamless Network will expand and enhance those relationships. The wider set of eventual national system partners such as other federal programs and state, tribal, and local government MPA sites and systems may benefit from this model. An active dialogue exists and will be maintained between the developing national system and the Seamless Network Initiative efforts in order to ensure that they complement one another.

Ocean Parks Stewardship Action Plan

The USOAP calls for the adoption of an Ocean Parks Strategy by the National Park Service. Key elements of this strategy include: characterizing marine species and habitats; evaluating and monitoring their condition; increasing the scientific understanding of how marine ecosystems function; and developing cooperative science-based fishery management plans between parks and state agencies. This plan was issued in December 2006 and can be viewed at http://www.nps.gov/pub_aff/oceans/Ocean_Park_ActionPlan.pdf. This important effort offers opportunities for collaborative approaches between the National Park Service, the Seamless Network initiative, and the national system to address shared science and management priorities.

C. International MPA Programs and Authorities

In addition to U.S. MPA programs and authorities, there are numerous international MPA efforts and linkages that can contribute to and benefit from the national system. Marine ecosystems and their associated natural resources rarely align with the political boundaries of sovereign countries. Moreover, ecosystems often overlap with adjacent countries and some natural resources may move back and forth between distant countries. In recognition of these important international connections, section 4(a) of the Order calls on federal agencies to identify opportunities to improve "linkages with, and technical assistance to, international [MPA] programs."

The United States shares a number of common resources with both neighboring and distant countries. For instance, migratory species (e.g., whales, sea turtles, pelagic fishes, and some birds) rely on the marine and coastal waters of multiple countries during various stages of their life. There are also a number of international law and policy issues regarding our underwater cultural heritage. For example, certain cultural resources that rest in the seabed of U.S. MPAs, such as sunken military craft and associated contents that have not been abandoned, retain their protected sovereign status and permanent right, title, and interest may be vested in the flag country.

Enhancing existing or establishing new linkages among systems in other countries can mutually benefit the United States and international MPAs through coordination of efforts, information and capacity sharing, and technical assistance. Along with sharing common resources, the United States also shares the consequences of potentially harmful activities occurring outside of U.S. waters, including pollution, over-harvesting of marine resources, and degradation of associated habitats. By coordinating with international MPA programs, the United States can minimize the harmful impacts of external activities and maximize the benefits of MPAs.

Table 4. Examples of Existing U.S. MPAs

MPA Name and Location	Name of Managing Entity and Type of Management	MPA Description*
Ashepoo-Combahee-Edisto (ACE) Basin National Estuarine Rese-arch Reserve South Carolina	Federal/State Partner-ship Management: National Oceanic and Atmospheric Admini-stration and South Carolina Depa-rtment of Natural Re-sources	ACE Basin is one of the largest undeveloped estuaries on the East Coast. Diverse estuarine wetlands provide an extensive complex of wildlife habitat types; the region contains 91,000 acres of tidal marshes, 26,000 acres of managed impoundments, and 12,000 acres of maritime islands.
Manele-Hulopoe Marine Life Conservation District (MLCD) Hawaii	State Management: Hawaii Department of Land and Natural Resources	The Manele-Hulopoe Marine Life Conservation District (MLCD) is located in the waters offshore of Palawai and Kamao on the southwestern coast of Lanai. Within Manele Bay corals are most abundant along the sides of the bay near the cliffs, where the bottom slopes off quickly to about 40 feet. The middle of the bay is a sand channel. Just outside the western edge of the bay near Pu'u Pehe rock, is "First Cathedrals," a popular SCUBA destination. Hulopo'e Bay has large tidepools at its left point. A shallow reef is just offshore, providing excellent snorkeling opportunities. Pu'u Pehe Cove has clear water and considerable marine life. Coral growth is interspersed with sand patches, and most coral is found away from the narrow beach in about 10 to 15 feet of water.
North Fork, St. Lucie Aquatic Preserve Florida	State Management: Florida Department of Environmental Protection	The North Fork, St. Lucie Aquatic Preserve contains various aquatic habitats such as riverine, blackwater stream, tidal marsh, slough, and floodplain forest communities. The headwaters of the North Fork are composed of freshwater from Ten Mile and Five Mile Creeks. Downstream, brackish conditions support tidal marshes with mangroves, leatherfern, and sawgrass.
Monomoy National Wildlife Refuge Massachusetts	Federal Management: Department of the Interior, U.S. Fish and Wildlife Service	Monomoy is comprised of 7,604 acres of barrier beach, sand dunes, freshwater ponds, and saltwater marshes. Monomoy provides habitat for hundreds of species of resting, feeding, and migratory birds. The refuge supports the largest nesting colony of common terns in the Gulf of Maine and second largest on the Atlantic Seaboard with close to 8,000 nesting pairs in 2001. Monomoy is the largest haul-out site of gray seals on the Atlantic Seaboard as well.

* Only the marine portion of the described areas are considered to be a part of the MPA; the terrestrial components, while a part of the larger management unit, are not considered to be part of the MPA.

For U.S. MPAs, important international linkages include, but are not limited to, those relating to Canada, Mexico, and Russia, as well as those amongst multiple countries in the Arctic, Pacific Islands, and Caribbean. Several legal mechanisms, such as bi-lateral and multi-lateral agreements and treaties, exist to address many of these resource management issues. For example, the International Maritime Organization's Particularly Sensitive Sea Areas program and the Wider Caribbean Protocol Concerning Specially Protected Areas and Wildlife are two MPA-related international efforts of significance. The MPA Center and/or its federal partners are actively involved in a number of such efforts, including the Commission on Environmental Cooperation's development of a North American MPA Network (NAMPAN) and the exchange of training and technical assistance with other nations. The national system can facilitate a dialogue and develop collaborative efforts between the United States and other countries to complement and support the work of MPA programs.

Appendix C. Programmatic Environmental Assessment

Lead Agency:
Department of Commerce, National Oceanic and Atmospheric Administration, National Ocean Service

Cooperating Agency:
Department of the Interior, National Park Service

Purpose and Need for This Programmatic Environmental Assessment

The purpose of this Programmatic Environmental Assessment (PEA) is to fulfill the requirements of Executive Order (EO) 13158, which are to develop, design and build a National System of Marine Protected Areas (MPAs).

Executive Order 13158 on MPAs

Executive Order 13158 on Marine Protected Areas (2000) calls on the Department of Commerce and the Department of the Interior (DOI), in consultation with other federal agencies and stakeholders, to develop a national system of marine protected areas (MPAs) to enhance the conservation of the nation's natural and cultural marine heritage. The Executive Order created the National Marine Protected Areas Center (MPA Center) within the National Oceanic and Atmospheric Administration (NOAA) to coordinate this effort. The mission of the MPA Center is to facilitate the effective use of science, technology, training, and information in the planning, management, and evaluation of the nation's system of marine protected areas.

The National System of MPAs

Currently, nearly 1,700 marine areas have been identified in the United States (U.S.). These areas are managed under the authority of hundreds of federal, state and territorial (state), tribal, and local laws and regulations. Familiar examples of MPAs include national and state marine sanctuaries, parks, wildlife refuges, and some fishery management areas. This patchwork of protected areas is an important component of the nation's marine conservation mission, but would be greatly enhanced by the improved coordination and integration across sites and MPA programs that a national system will provide.

The National System of MPAs (national system) will be built collaboratively by existing MPA sites and systems through partnerships at the ecosystem, regional, and national levels. The national system will focus on supporting shared priorities for enhancing coordination and stewardship of partner MPA sites and systems in order to improve effectiveness. The national system may ultimately include some new areas vital to the conservation of significant natural and cultural marine resources. These may be identified by national system partners through regional planning or other processes, and will be based on the best available science and stakeholder involvement.

Any new MPAs would need to be designated through an existing federal, state, tribal, or local authority, as the Executive Order provides no authority to create new MPAs.

Need for Action

The Executive Order calls on the MPA Center to develop a Framework for the national system (Framework). The first draft was published for public comment in September 2006, and was revised after due consideration of comments and recommendations received. A second draft was published for an additional round of public comment during March-May 2008, and again has been revised with consideration of input received. This PEA has also been revised based on comments received during the 2006-2008 comment period.

The purpose of the Framework is to serve as a "road map" for developing the national system that will specify a common vision, and common goals, objectives, and criteria for the national system, as well as the process for partnerships among federal, state, tribal, and local government agencies and stakeholders to develop it. While the Executive Order and the Framework document are non-regulatory, the MPA Center developed this PEA to provide federal and state agencies, tribes, and other stakeholders with the best available information on the potential impacts of the Framework document during its two public comment periods.

Scope of this Analysis

This PEA considers the programmatic environmental consequences of proposing the Framework. As previously described, the Framework itself only lays out a strategic process to achieve a national system of MPAs. The Framework itself does not propose any new MPAs, nor does it create or recommend any new authority under which they may be designated.

The consideration of designating additional MPAs or expanding existing MPAs will occur solely at the discretion of the state, federal, tribal, and local agencies which have the authority to develop different MPAs to fulfill their own missions and implement the national system. As such, any potential site-specific environmental, economic, and social impacts cannot be meaningfully analyzed until these agencies consider individual MPA proposals under their own authorities. Therefore, the potential effects of any detailed regional, state, or local MPA alternatives proposed by a federal agency under this Framework would be further analyzed under NEPA at the time they are proposed, including in environmental assessments tiered from this PEA as appropriate.

Description of Proposed Action and Alternatives

Alternatives Considered, but Rejected

In considering alternatives for proposing the Framework, the following three were selected as constituting a reasonable range of alternatives for this PEA: "Alternative A: Take No Action," "Alternative B: Propose the Draft Framework for Developing the National System of Marine Protected Areas" and "Alternative C: Propose the Framework for the National System of Marine Protected Areas of the United States of America." Numerous other possible alternatives were, however, considered by NOAA for analysis, but ultimately rejected. For example, a wide range of alternatives would have resulted from all the possible permutations of changes in the Framework's approach to meeting the various requirements of the MPA Executive Order. Several factors led to the determination that the approach of analyzing a wide range of many potential alternatives should be rejected.

First, the Framework lays out a series of processes for U.S. MPA programs, managing entities, authorities, and other stakeholders around the country to work together to determine eligible MPAs and the most appropriate, specific approaches for developing the national system. Because the Framework is focused on managing entity and stakeholder processes to determine specific approaches and actions, the environmental consequences of these permutations cannot be predicted to be significantly different than Alternative C.

Second, and most important, the processes outlined in the elements of the Framework are based on input received from consultations with and recommendations from MPA stakeholders around the country, including the MPA Federal Advisory Committee, as required by the Executive Order. Creating a range of alternatives that are either independent of these consultations or consider only some of the recommendations received would not meet the requirements of the Executive Order.

Therefore, having considered additional alternatives for proposing the Framework for the national system, NOAA has determined that the three alternatives described below constitute a reasonable and practical range of alternatives for assessing the anticipated environmental consequences of fulfilling the need to develop the Framework.

Alternative A: Take No Action

Under this alternative, NOAA would not propose a Framework as required by the MPA Executive Order. The MPA Executive Order would stand alone without any further detail of the processes necessary for developing the national system. There would be no description of processes for identifying and including existing MPAs in the national system, working with MPA programs to collaboratively identify and address common stewardship needs, or identifying place-based gaps in protection.

Alternative B: Propose the Draft Framework for Developing the National System of Marine Protected Areas

NOAA proposed the first draft of the Framework published in September 2006. As noted by the MPA Federal Advisory Committee and many public comments, this draft document lacked a strategic focus to describe how the national system would target priority conservation objectives; lacked design and implementation principles to guide development of the system; and provided only a minimal description of how the national system would be coordinated and conduct gap analyses on a regional basis.

Alternative C: Propose the Framework for the National System of Marine Protected Areas of the United States of America (Preferred)

This alternative would fulfill the directive of the MPA Executive Order to develop a Framework. The Framework provides guidance for developing the national system and therein implementing key elements of the Executive Order. The full descriptions of the proposed national system elements and associated processes are contained in the Framework and summarized here as:

- Summary of authority for developing the Framework and national system.
- Overview of key U.S. MPA programs and related initiatives.
- Key definitions for developing the national system.
- Goals and objectives for the national system.
- Sequence and steps for implementing the Framework.
- Process for identifying, nominating, and including MPAs in the national system.
- Options for building collaborative efforts to enhance stewardship and regional coordination of MPAs.
- Process for identifying conservation gaps in the national system.
- Maintenance of the official List of National System MPAs.
- Process for implementing the "avoid harm" provision.
- Options for evaluating effectiveness of the national system.
- Mechanisms for tracking and reporting national system progress and priorities.

Description of Affected Environment

The geographic extent of the Framework and the nation's existing MPAs that it aims to support span the United States' territorial waters and Exclusive Economic Zone waters of the Pacific Ocean, including the Bering Sea; Atlantic Ocean, including the Gulf of Mexico and Caribbean Sea; Arctic Ocean; and the Great Lakes. This environment encompasses the entire range of the nation's marine ecosystems, including their natural heritage, cultural heritage, and sustainable production resources and functions, goods, and services. The following are general descriptions of five valued environmental components that may be affected programmatically by the Framework. More detailed descriptions of specific affected environments will be given in future tiered analyses based on future consideration of MPAs which may occur under the authority of individual state, federal, tribal, and local agencies.

Natural Heritage Resources

The nation's existing MPAs, whether managed by federal, state, tribal, or an inter-governmental collaboration of entities help to conserve and restore the wealth of U.S. natural marine environments, including but not limited to, kelp forests, warm and cold water coral reefs, rocky intertidal areas, offshore banks and seamounts, estuarine areas, the Great Lakes waters, deep sea vents, and sand and mud flats. In these marine environments, MPAs play an important role in protecting the significant natural biological communities, endangered and threatened species, habitats, ecosystems, processes, and the ecological services, uses, and values they provide to this and future generations. These various components of the nation's marine environment are critical to maintaining the integrity and health of marine and coastal ecosystems. Oftentimes managing for one of these elements means protecting the others. For example, to effectively manage endangered or threatened species, the habitat they rely upon must also be protected.

Sustainable Production Resources

Existing U.S. MPAs are also designed and established with the intent to help ensure the sustainability of the renewable living resources and their habitats, including, but not limited to, spawning, mating, and nursery grounds, and areas established to minimize bycatch of species that are important to the nation's economy and the livelihoods and subsistence needs of its citizens. MPAs can help to sustain commercial and recreational fisheries by controlling fishing effort, protecting critical stages in the life history of fishery species, conserving genetic diversity of exploited species, reducing secondary impacts of fishing on essential fish habitat and other species, and ensuring against fisheries collapse (Murray et al. 1999; NRC, 2001). MPAs may allow site-specific regulation of selected species, selected gear types, or fishing methods. Certain MPAs or zones within MPAs may be fishery reserves that protect all or nearly all species from fishing. Many studies indicate that abundance and size of target species increase in marine protected areas that limit extractive use (Dugan and Davis, 1993; Crowder et al., 2000; Halpern, 2003).

Cultural Heritage Resources

The nation's existing MPAs preserve and protect important cultural resources. These cultural resources reflect the nation's maritime history and traditional cultural connections to the sea, as well as the uses and values they provide to this and future generations. Examples include archeological sites that contain significant cultural artifacts; sunken historic ships, aircraft, or other vessels; and areas important to specific cultures. Protecting cultural resources in MPAs reduces the chance that artifacts will be removed or damaged from modern-day commercial or recreational activities. Unlike many biological communities that have some level of resilience to recover from degradation, once cultural sites are damaged, the information and value of these non-renewable resources may be lost forever. MPAs are an important tool for conserving cultural resources by monitoring the environment for change and stabilizing deteriorating structures. MPAs also encourage actions to find, preserve, and interpret the associated artifacts that may otherwise be inaccessible to the public. By protecting marine sites that are important to the nation's diverse cultures, existing U.S. MPAs preserve a part of history for future generations.

Current Governmental Management Structure

The past several decades have witnessed a dramatic increase in the use of MPAs as a conservation and management tool to protect the nation's most important natural and cultural marine resources and areas. Over 90 percent of U.S. MPAs were established after 1970 (National MPA Center Marine Protected Area Inventory, 2008). The growth in MPAs has not only resulted in increased protections to certain natural and cultural marine resources, but also brought about a significant number of new MPA programs and authorities at all levels of government, each with their own requirements, levels of protection, and associated terms.

These programs and the MPA sites that they manage are components of a complex sociopolitical landscape that features diverse institutions, governance structures, and processes. They include, for example, federal programs such as the National Marine Sanctuaries and National Parks; tribal MPA authorities and co-management arrangements with states; state programs such as fish and wildlife, coastal zone management, and historic preservation; and other governmental approaches to MPAs.

Each of these programs has its own mandate it is required to fulfill. These mandates often overlap in both geographic scope and the conservation purposes for which they were established. In addition, while many existing MPA programs comprise a system of MPAs, there are a limited number of mechanisms in place to coordinate MPA efforts across ecosystem, regional, national, or international levels among MPA programs and levels of government. This is not to say that no such coordination is happening. In fact, there are a number of good examples of existing MPA sites and programs in a common geography working together, which serve as excellent models. However, there is no overarching MPA framework for facilitating and promoting such coordination across levels of government and at an ecosystem or regional scale around the nation. Similarly, the effectiveness of the existing suite of MPAs in contributing to the long-term sustainability of important resources, habitats and ecosystems, and the services and values they provide, is largely yet to be determined.

Social, Economic and Cultural Benefits

MPAs in the United States and its territories provide social, economic, and cultural benefits by protecting resources and environments. These benefits come in many forms, both tangible and intangible and direct and indirect. Direct, tangible benefits may include supporting the socioeconomic well-being of communities tied to our nation's fisheries by enhancing stocks for sustainable harvest and recreational opportunities. These communities provide significant inputs to the U.S. economy and many have long and storied historical connections to the marine environment. MPAs that ensure sustainable production have the intangible benefit of promoting cultural continuity and identity, which is instrumental in maintaining healthy communities.

By protecting key resources and habitats, MPAs can also promote greater economic returns from tourism through enhanced visitor experiences. These direct economic benefits are inextricably linked with the intangible quality of visitor experience. Good water quality, abundant living resources, and scenic, aesthetic ocean environments attract visitors to coastal areas around the globe. These visitors engage in diverse activities that include non-extractive uses of the marine environment, such as scuba diving, snorkeling, wildlife watching, boating, and surfing, as well as extractive uses such as fishing. All of these activities rely on healthy marine environments. U.S. MPAs help ensure that marine environments will continue to draw the visitors that have become critical to many coastal economies. For example, in Monroe County, Florida, location of the Florida Keys National Marine Sanctuary and other marine-related parks and wildlife refuges, the estimated total tourist contribution to the economy (1995-1996) is over 60 percent (English et al., 1996).

MPAs also provide direct, tangible benefits by providing opportunities for research and education. Certain MPAs feature academic and applied monitoring of short-term events and long-term environmental trends, as well as biomedical research (Salm et al, 2000).

MPAs can provide hands-on experience and outdoor laboratories for bringing classroom studies to life. MPA educational programs have the potential to promote public awareness of the importance of marine ecosystems and their many benefits.

MPAs also protect historic connections to our nation's heritage that are critical to social and cultural continuity. People and communities are connected to marine resources, including both natural and cultural features. These connections are affirmed through direct practice, oral and written narrative, and everyday discourse. MPAs can enhance cultural connectivity to places by ensuring their protection for future generations, allowing traditional cultural practices, promoting awareness of our nation's heritage, and acknowledging existence and bequest values inherent in marine resources.

Environmental Consequences of Proposed Action and Alternatives

As previously noted, the Framework only provides a strategic process for establishing the National System of MPAs, rather than proposing any specific action itself. Therefore, at a programmatic level, the environmental consequences of the proposed action and alternatives are negligible. The specific environmental, economic, social, and cumulative impacts of proposed new or expanded MPAs later proposed by a federal agency under this Framework would be further analyzed under NEPA at the time they are proposed, including in environmental assessments tiered from this PEA as appropriate.

Alternative A: Take No Action

Environmental Impacts

Taking no action would result in no predictable direct or indirect environmental impacts, either positive or negative. The 'Take No Action' alternative would not allow for the realization of the benefits expected from the proposed Framework's greater integration and coordination of conservation efforts among existing authorities and sites.

Socioeconomic Impacts

Taking no action would result in no predictable direct socioeconomic impacts, either positive or negative. The 'Take No Action' alternative would not allow for the realization of the benefits expected from the proposed Framework's greater integration and coordination of conservation efforts among existing authorities and sites.

Alternative B: Propose the Draft Framework for Developing the National System of Marine Protected Areas

Environmental Impacts

The Draft Framework would not be expected to result in adverse impacts on the environment. The Draft Framework proposed to coordinate the activities among federal, state, tribal, and local MPA sites and systems to reduce administrative costs and promote efficiency and the effective use of existing management infrastructure for marine resource protection. However, because of the lack of a strategic focus within this alternative, the expected beneficial long-term environmental impacts and improved quality of the nation's marine resources would not be as great as those under Alternative C.

Socioeconomic Impacts

Similar to Alternative C, the proposed Draft Framework would not be expected to result in adverse socioeconomic impacts. However, because of the lack of focused design and implementation principles, and a clear vision for regional coordination, there is less potential, relative to Alternative C, for long-term positive socioeconomic impacts from promoting integration among government authorities, enhancing knowledge and awareness of MPAs as a tool of ecosystem-based management, and supporting processes for incorporating stakeholders and communities in ecosystem management.

Alternative C: Propose the Framework for the National System of Marine Protected Areas of the United States of America (Preferred)

Environmental Impacts

The Framework is not expected to result in adverse impacts on the environment. The Framework proposes to coordinate the activities among federal, state, tribal, and local MPA sites and systems to reduce administrative costs and promote efficiency and the effective use of existing management infrastructure for marine resource protection.

Implementation of the Framework provides opportunities for shared information, resources, scientific expertise, and lessons learned for individual MPAs. The proposed Framework mostly involves a number of low or no impact activities that will positively affect the stewardship and management of individual MPAs and ultimately lead to beneficial long-term environmental impacts and improved quality of the nation's marine resources relative to Alternative A. Additional environmental analysis of future activities, as required under the

National Environmental Policy Act (NEPA) and other acts and executive orders, would be prepared as necessary by the relevant entity or entities taking any such actions.

The Framework also promotes activities over time to identify gaps in protection of important marine resources and subsequent area-based conservation priorities that would be needed to manage and protect those resources. This component of the Framework is similarly comprised of a number of low or no impact activities that ultimately could lead to beneficial long-term environmental impacts relative to Alternative A. In order to realize these benefits, however, actions to implement new or increased protections would be needed. Activities taken by individual entities in the future, such as changes in MPA regulations or the establishment of new MPAs as a result of the implementation of the proposed Framework will undergo separate NEPA analysis by entities taking such actions as required and appropriate.

Socioeconomic Impacts

The proposed Framework is not expected to result in adverse socioeconomic impacts. The Framework provides guidance for the implementation of the national system. It does not establish new MPAs or directly affect the stewardship and management, including human uses and values, associated with existing MPAs. The socioeconomic impacts of, for example, the long-term cumulative effects of developing the national system will be assessed as necessary under NEPA and other federal mandates for specific actions taken by those entities or programs with the authority to establish and manage MPAs and/or alter MPA regulations.

In proposing to integrate the activities and conservation objectives among the various authorities, the Framework will have its most immediate effects upon the communication and organizational structures across the various levels of MPA governance. As a result, there is great potential, relative to Alternative A, for long-term positive socioeconomic impacts from promoting integration among government authorities, enhancing knowledge and awareness of MPAs as a tool of ecosystem-based management, and supporting processes for incorporating stakeholders and communities in ecosystem management.

Furthermore, the implementation of the national system as proposed by the Framework will have long-term positive impacts, relative to Alternative A, for participating MPA sites, their associated marine resources, and the wider ecosystems of which they are a part. The national system will seek to integrate natural heritage, cultural heritage, and sustainable production objectives in order to minimize adverse socioeconomic impacts and promote comprehensive MPA conservation and management. It will focus on improving the effectiveness of MPA design, management, and evaluation through dissemination and use of the best available science and tools.

Additional socioeconomic analysis as required under NEPA and other acts and executive orders, would be prepared by the relevant entity or entities as necessary for future specific actions.

Cumulative Effects

The Council on Environmental Quality regulations (40 CFR 1508.8) define cumulative effects as "impacts on the environment that result from the incremental impact of the action when added to other past, present, and reasonably foreseeable future actions regardless of what agency or person undertakes such actions." At a programmatic level, the integration and coordination of federal, state, local and tribal agencies to improve MPA conservation and management are anticipated to have no significant adverse cumulative impact to environmental or socioeconomic resources. Relative to Alternative A, the proposed action has beneficial cumulative impacts to the resources that the National System of MPAs will protect. At a programmatic level, socioeconomic impacts are anticipated to be negligible (see above). Future tiered analyses on specific alternatives and resources will occur as entities consider future actions which fall under this Framework.

REFERENCES

Crowder, L. B, S. J., Lyman, W.F. & Figueira, J. Priddy. (2000). Source-sink population dynamics and the problem of siting marine reserves. *Bulletin of Marine Science 66(3)*: 799-820.

Dugan, J. E. & G. E. Davis. (1993). Applications of marine refugia to coastal fisheries management. *Canadian Journal of Fisheries and Aquatic Science 50*:2029-2042.

English, D. B. K., W. Kriesel, V. R. Leeworthy, & P. C. Wiley. (1996). *Economic Contribution of Recreating Visitors to the Florida Keys/Key West. National Oceanic and Atmospheric Administration*. Silver Spring, MD. I + 22pp.

Halpern, B. S. (2003). The impact of marine reserves: Do reserves work and does reserve size matter? *Ecological Applications 13(1) Supplement*: S117-S137.

Nuclear Regulatory Commission. (2001). *Marine protected areas: Tools for sustaining ocean ecosystems*. National Academy Press, Washington, DC.

Murray, S. N., R. F. Ambrose, J. A. Bohnsack, L. W. Botsford, M. H. Carr, G. E. Davis, P. K. Dayton, D. Gotshall, D. R. Gunderson, M. A. Hixon, J. Lubchenco, M. Mangel, A. MacCall, D. A. McArdle, J. C. Ogden, J. Roughgarden, R. M. Starr, M. J., & Tegner, M. M. Yoklavich. (1999). No-take reserve networks: *Protection for fishery populations and marine ecosystems. Fisheries 24(11)*:11-25.

Salm, R. V., J. Clark, & E. Siirila. (2000). *Marine and Coastal Protected Areas*: A Guide for Planners and Managers. IUCN. Washington, DC. xxi + 371pp.

Finding of No Significant Impact

The Council on Environmental Quality (CEQ) Regulations state that the determination of significance using an analysis of effects requires examination of both context and intensity, and lists ten criteria for intensity (40 CFR 1508.27). In addition, the National Oceanic and Atmospheric Administration Administrative Order (NAO) 216-6 Section 6.01b. 1 - 11 provides eleven criteria, including the same ten as the CEQ Regulations and one additional, for determining whether the impacts of a proposed action are significant. Each criterion is discussed below with respect to the proposed action and considered individually as well as in combination with the others.

1. **Can the proposed action reasonably be expected to cause both beneficial and adverse impacts that overall may result in a significant effect, even if the effect will be beneficial?**
   NOAA expects the implementation of the proposed Framework will result in a number of activities that will positively affect the stewardship and management of individual MPAs and ultimately lead to beneficial long-term environmental impacts and improved quality of the nation's marine resources. The specific environmental, economic, social, and cumulative impacts of any proposed new or expanded MPAs later proposed by a federal agency under this Framework would be further analyzed as required by NEPA at the time they are proposed.

2. **Can the proposed action reasonably be expected to significantly affect public health or safety?**
   No negative impacts to public health or safety are associated with these activities.

3. **Can the proposed action reasonably be expected to result in significant impacts to unique characteristics of the geographic area, such as proximity to historic or cultural resources, park lands, prime farmlands, wetlands, wild and scenic rivers, or ecologically critical areas?**
   The Framework for the national system will not have significant adverse impacts on the areas listed above. It will provide a mechanism for coordination among existing marine protected areas, including those that protect significant natural and cultural marine resources. The Framework is expected to enhance the effectiveness of participating MPAs in contributing to national conservation objectives, such as the protection of spawning and nursery areas or the conservation of resources listed on the National Register of Historic Places.

4. **Are the proposed action's effects on the quality of the human environment likely to be highly controversial?**
   While individual MPAs are often a contentious subject, the effects of the proposed Framework on the human environment are not likely to be controversial. The actions and activities associated with the various components of the Framework focus on promoting coordination, collaboration, opportunities for stakeholder input, and enhancing scientific understanding in support of the effective use of MPAs. These activities largely have little or no impact on the human environment, but are envisioned to positively affect the stewardship and management of individual MPAs and ultimately lead to beneficial long-term impacts on the human environment and improved quality of the nation's marine resources.

5. **Are the proposed action's effects on the human environment likely to be highly uncertain or involve unique or unknown risks?**
   The Framework's effects are not expected to involve unique or unknown risks. Work will focus on enhancing coordination; sharing best management practices, technologies and science; and establishing conservation partnerships across all levels of government and with stakeholders.

6. **Can the proposed action reasonably be expected to establish a precedent for future actions with significant effects or represent a decision in principle about a future consideration?**
The Framework does not establish a precedent for future actions with significant effects. Regional conservation gap analyses will identify ecologically and culturally significant areas that may require additional protection. However, the Framework does not have any authority to establish a new MPA or another type of protection for these areas. Any additional protection would be provided under existing federal, state, local or tribal laws, and would be subject to the required review processes under the respective authority.

7. **Is the proposed action related to other actions that when considered together will have individually insignificant but cumulatively significant impacts?**
The activities associated with the proposed Framework largely have little or no impact on the human environment, but are envisioned to positively affect the stewardship and management of individual MPAs and ultimately lead to beneficial long-term impacts on the human environment and improved quality of the nation's marine resources. By providing the first national geospatial database of MPAs across all levels of government, the national system will provide an opportunity to better understand the cumulative effectiveness of existing MPAs and to identify opportunities for collaboration. The cumulative effects of specific MPAs that may be proposed under the Framework will be analyzed in the NEPA analysis prepared for that proposed action.

8. **Can the proposed action reasonably be expected to adversely affect districts, sites, highways, structures, or objects listed in or eligible for listing in the National Register of Historic Places or may cause loss or destruction of significant scientific, cultural, or historical resources?**
The Framework will not adversely affect any of the aforementioned areas. It will benefit significant scientific, cultural and historical resources and areas listed in or eligible for listing in the National Register of Historic Places, as the protection of these areas is included in the goals and objectives of the national system.

9. **Can the proposed action reasonably be expected to have a significant impact on endangered or threatened species, or their critical habitat as defined under the Endangered Species Act of 1973?**
The Framework will not adversely affect endangered or threatened species or their critical habitat. The conservation of critical habitat for threatened and endangered species is an objective of the national system. The national system will provide tools for analyzing and mapping existing protected areas that contribute to the conservation of threatened and endangered species, as well as gaps in the protection of critical habitat where new MPAs may be needed.

10. **Can the proposed action reasonably be expected to threaten a violation of Federal, state, or local law or requirements imposed for environmental protection?**

The Framework will not threaten any violation of Federal, state, or local law or requirements for environmental protection.

11. **Can the proposed action reasonably be expected to result in the introduction or spread of a nonindigenous species?**
    The Framework will not result in the introduction or spread of any nonindigenous species. By providing a mechanism for regional coordination, it will help MPAs develop shared strategies and partnerships to prevent and contain the impacts of nonindigenous species.

List of Preparers and Agencies Consulted

Department of Commerce, National Oceanic and Atmospheric Administration
Department of the Interior, National Park Service

APPENDIX D. EXECUTIVE ORDER 13158

Executive Order 13158
Presidential Documents
Executive Order 13158 of May 26, 2000
Marine Protected Areas

By the authority vested in me as President by the Constitution and the laws of the United States of America and in furtherance of the purposes of the National Marine Sanctuaries Act (16 U.S.C. 1431 et seq.), National Wildlife Refuge System Administration Act of 1966 (16 U.S.C. 668dd-ee), National Park Service Organic Act (16 U.S.C. 1 et seq.), National Historic Preservation Act (16 U.S.C. 470 et seq.), Wilderness Act (16 U.S.C. 1131 et seq.), Magnuson-Stevens Fishery Conservation and Management Act (16 U.S.C. 1801 et seq.), Coastal Zone Management Act (16 U.S.C. 1451 et seq.), Endangered Species Act of 1973 (16 U.S.C. 1531 et seq.), Marine Mammal Protection Act (16 U.S.C. 1362 et seq.), Clean Water Act of 1977 (33 U.S.C. 1251 et seq.), National Environmental Policy Act, as amended (42 U.S.C. 4321 et seq.), Outer Continental Shelf Lands Act (42 U.S.C. 1331 et seq.), and other pertinent statutes, it is ordered as follows:

Section 1. Purpose. This Executive Order will help protect the significant natural and cultural resources within the marine environment for the benefit of present and future generations by strengthening and expanding the Nation's system of marine protected areas (MPAs). An expanded and strengthened comprehensive system of marine protected areas throughout the marine environment would enhance the conservation of our Nation's natural and cultural marine heritage and the ecologically and economically sustainable use of the marine environment for future generations. To this end, the purpose of this order is to, consistent with domestic and international law: (a) strengthen the management, protection, and conservation of existing marine protected areas and establish new or expanded MPAs; (b) develop a scientifically based, comprehensive national system of MPAs representing diverse

U.S. marine ecosystems, and the Nation's natural and cultural resources; and (c) avoid causing harm to MPAs through federally conducted, approved, or funded activities.

Sec. 2. Definitions. For the purposes of this order:

a. "Marine protected area" means any area of the marine environment that has been reserved by Federal, State, territorial, tribal, or local laws or regulations to provide lasting protection for part or all of the natural and cultural resources therein.
b. "Marine environment" means those areas of coastal and ocean waters, the Great Lakes and their connecting waters, and submerged lands thereunder, over which the United States exercises jurisdiction, consistent with international law.
c. The term "United States" includes the several States, the District of Columbia, the Commonwealth of Puerto Rico, the Virgin Islands of the United States, American Samoa, Guam, and the Commonwealth of the Northern Mariana Islands.

Sec. 3. MPA Establishment, Protection, and Management. Each Federal agency whose authorities provide for the establishment or management of MPAs shall take appropriate actions to enhance or expand protection of existing MPAs and establish or recommend, as appropriate, new MPAs. Agencies implementing this section shall consult with the agencies identified in subsection 4(a) of this order, consistent with existing requirements.

Sec. 4. National System of MPAs. (a) To the extent permitted by law and subject to the availability of appropriations, the Department of Commerce and the Department of the Interior, in consultation with the Department of Defense, the Department of State, the United States Agency for International Development, the Department of Transportation, the Environmental Protection Agency, the National Science Foundation, and other pertinent Federal agencies shall develop a national system of MPAs. They shall coordinate and share information, tools, and strategies, and provide guidance to enable and encourage the use of the following in the exercise of each agency's respective authorities to further enhance and expand protection of existing MPAs and to establish or recommend new MPAs, as appropriate:

- science-based identification and prioritization of natural and cultural resources for additional protection;
- integrated assessments of ecological linkages among MPAs, including ecological reserves in which consumptive uses of resources are prohibited, to provide synergistic benefits;
- a biological assessment of the minimum area where consumptive uses would be prohibited that is necessary to preserve representative habitats in different geographic areas of the marine environment;
- an assessment of threats and gaps in levels of protection currently afforded to natural and cultural resources, as appropriate;
- practical, science-based criteria and protocols for monitoring and evaluating the effectiveness of MPAs;

- identification of emerging threats and user conflicts affecting MPAs and appropriate, practical, and equitable management solutions, including effective enforcement strategies, to eliminate or reduce such threats and conflicts;
- assessment of the economic effects of the preferred management solutions; and
- identification of opportunities to improve linkages with, and technical assistance to, international marine protected area programs.

b. In carrying out the requirements of section 4 of this order, the Department of Commerce and the Department of the Interior shall consult with those States that contain portions of the marine environment, the Commonwealth of Puerto Rico, the Virgin Islands of the United States, American Samoa, Guam, and the Commonwealth of the Northern Mariana Islands, tribes, Regional Fishery Management Councils, and other entities, as appropriate, to promote coordination of Federal, State, territorial, and tribal actions to establish and manage MPAs.

c. In carrying out the requirements of this section, the Department of Commerce and the Department of the Interior shall seek the expert advice and recommendations of non-Federal scientists, resource managers, and other interested persons and organizations through a Marine Protected Areas Federal Advisory Committee. The Committee shall be established by the Department of Commerce.

d. The Secretary of Commerce and the Secretary of the Interior shall establish and jointly manage a website for information on MPAs and Federal agency reports required by this order. They shall also publish and maintain a list of MPAs that meet the definition of MPA for the purposes of this order.

e. The Department of Commerce's National Oceanic and Atmospheric Administration shall establish a Marine Protected Area Center to carry out, in cooperation with the Department of the Interior, the requirements of subsection 4(a) of this order, coordinate the website established pursuant to subsection 4(d) of this order, and partner with governmental and nongovernmental entities to conduct necessary research, analysis, and exploration. The goal of the MPA Center shall be, in cooperation with the Department of the Interior, to develop a framework for a national system of MPAs, and to provide Federal, State, territorial, tribal, and local governments with the information, technologies, and strategies to support the system. This national system framework and the work of the MPA Center is intended to support, not interfere with, agencies' independent exercise of their own existing authorities.

f. To better protect beaches, coasts, and the marine environment from pollution, the Environmental Protection Agency (EPA), relying upon existing Clean Water Act authorities, shall expeditiously propose new science-based regulations, as necessary, to ensure appropriate levels of protection for the marine environment. Such regulations may include the identification of areas that warrant additional pollution protections and the enhancement of marine water quality standards. The EPA shall consult with the Federal agencies identified in subsection 4(a) of this order, States, territories, tribes, and the public in the development of such new regulations.

Sec. 5. Agency Responsibilities. Each Federal agency whose actions affect the natural or cultural resources that are protected by an MPA shall identify such actions. To the extent permitted by law and to the maximum extent practicable, each Federal agency, in taking such actions, shall avoid harm to the natural and cultural resources that are protected by an MPA. In implementing this section, each Federal agency shall refer to the MPAs identified under subsection 4(d) of this order.

Sec. 6. Accountability. Each Federal agency that is required to take actions under this order shall prepare and make public annually a concise description of actions taken by it in the previous year to implement the order, including a description of written comments by any person or organization stating that the agency has not complied with this order and a response to such comments by the agency.

Sec. 7. International Law. Federal agencies taking actions pursuant to this Executive Order must act in accordance with international law and with Presidential Proclamation 5928 of December 27, 1988, on the Territorial Sea of the United States of America, Presidential Proclamation 5030 of March 10, 1983, on the Exclusive Economic Zone of the United States of America, and Presidential Proclamation 7219 of September 2, 1999, on the Contiguous Zone of the United States.

Sec. 8. General.

a. Nothing in this order shall be construed as altering existing authorities regarding the establishment of Federal MPAs in areas of the marine environment subject to the jurisdiction and control of States, the District of Columbia, the Commonwealth of Puerto Rico, the Virgin Islands of the United States, American Samoa, Guam, the Commonwealth of the Northern Mariana Islands, and Indian tribes.
b. This order does not diminish, affect, or abrogate Indian treaty rights or United States trust responsibilities to Indian tribes.
c. This order does not create any right or benefit, substantive or procedural, enforceable in law or equity by a party against the United States, its agencies, its officers, or any person.

(Presidential Sig.) William J. Clinton
THE WHITE HOUSE,
May 26, 2000.

Appendix E. MPA FAC and Ex Officio Members, and the Federal MPA Interagency Working Group

Current Members of the MPA Federal Advisory Committee

Chair

Dr. Mark Hixon, Professor, Department of Zoology, Oregon State University

Vice-Chair

Mr. Robert Zales, II, Owner, Bob Zales Charters

Members

Ms. Lori Arguelles, President and CEO, National Marine Sanctuaries Foundation
Mr. Charles D. Beeker, Director, Office of Underwater Science, School of Health, Physical Education and Recreation, Indiana University
Mr. David Benton, Benton & Associates
Dr. Daniel Bromley, Professor, Department of Agricultural and Applied Economics, University of Wisconsin
Dr. Anthony Chatwin, Marine Conservation Planner, The Nature Conservancy
Mr. Rick Gaffney, Pacific Boats and Yachts
Dr. Steve Gaines, Professor, Ecology, Evolution and Marine Biology, University of California, Santa Barbara
Ms. Ellen Goethel, Co-Owner, "Ellen Diane" / Ocean Educator
Dr. Dennis Heinemann, Senior Scientist, The Ocean Conservancy
Mr. George Lapointe, Commissioner, Maine Department of Marine Resources
Mr. Victor T. Mastone, Director and Chief Archeologist, Massachusetts Board of Underwater Archaeological Resources
Ms. Melissa Miller-Henson, Program Manager, California Marine Life Protection Act Initiative
Dr. Russell Moll, Director, California Sea Grant College Program, University of California, San Diego
Dr. Elliott Norse, President, Marine Conservation Biology Institute
Dr. John Ogden, Director and Professor, Florida Institute of Oceanography, University of South Florida
Mr. Terry O'Halloran, Hawaii Superferry, Tourism Business Solutions, LLC
Mr. Alvin D. Osterback, Port Director, City of Unalaska/Port of Dutch Harbor
Dr. Walter Pereyra, Chairman, Arctic Storm Management Group, Inc.
Mr. Eugenio Piñeiro-Soler, Chairman, Caribbean Fishery Management Council
Dr. Robert S. Pomeroy, Sea Grant Fisheries Specialist, Connecticut Sea Grant Office, University of Connecticut at Avery Point
Mr. Gilbert Radonski, Fisheries Consultant, Former President, Sport Fishing Institute
Mr. James P. Ray, President, Oceanic Environmental Solutions, LLC
Captain Philip G. Renaud, USN (Ret.), Executive Director, Living Oceans Foundation
Mr. Jesús C. Ruiz, President, California Divers
Mr. Bruce A. Tackett, Manager, Legislative and Regulatory Issues, ExxonMobil Biomedical Sciences, Inc.
Mr. David H. Wallace, Owner, Wallace and Associates
Mr. Robert Wargo, President, North American Submarine Cable Association, Marine Liaison Manager, AT&T

Ex Officio Federal Representatives

Department of Commerce

Ms. Laura Furgione, Assistant Administrator for Program Planning and Integration, NOAA

Department of the Interior

Dr. Kaush Arha, Deputy Assistant Secretary for Fish and Wildlife and Parks, U.S. Department of the Interior
Designee: Mr. Randal Bowman, Office of the Assistant Secretary, Fish and Wildlife and Parks, U.S. Department of the Interior

Department of Defense/Navy

Mr. Donald Schregardus, Deputy Assistant Secretary of the Navy (Environment)
Designee: Capt. Robin Brake, Director, Marine Science, Office of the Assistant Secretary of the Navy (Installations and Environment)

Department of Defense/Army Corps

Mr. Joseph Wilson, U.S. Army Corps of Engineers, South Atlantic Division

Department of Homeland Security

Rear Admiral Wayne Justice, Assistant Commandant for Response, U.S. Coast Guard
Designee: LCDR Chris Barrows, Commandant (CG-3RPL-4), Chief, Fisheries and Marine Protected Species Law Enforcement, US Coast Guard

U.S. Agency for International Development

Ms. Jacqueline Schafer, Deputy Assistant Administrator, Bureau for Economic Growth, Agriculture and Trade
Designee: Dr. Barbara Best, Coastal Resources and Policy Advisor, Office of Natural Resources Management, Bureau for Economic Growth, Agriculture and Trade

National Science Foundation

Ms. Roxanne Nikolaus, Ocean Sciences Division

Department of Agriculture

Mr. Merlin Bartz, Office of the Under Secretary for Conservation, Natural Resources and the Environment

Department of State

Ms. Margaret F. Hayes, Director of the Office of Oceans Affairs, Bureau of Oceans and International Environmental and Scientific Affairs

Environmental Protection Agency

Dr. Brian Melzian, Oceanographer/Project Officer, U.S. Environmental Protection Agency, National Health and Environmental Effects Research Laboratory, Atlantic Ecology Division

Designated Federal Official

Ms. Lauren Wenzel, NOAA Ocean Service, National Marine Protected Area Center

Past Members of the Mpa fac

Members

Dr. Tundi Agardy, Executive Director, Sound Seas
Mr. Robert Bendick, Jr., Vice President, Southeast Division, The Nature Conservancy
Dr. Michael Cruickshank, President, Marine Minerals Technology Center Associates
Ms. Carol Dinkins, Partner, Vinson and Elkins Attorneys at Law
Dr. Rodney Fujita, Senior Scientist, Environmental Defense
Dr. Delores Garza, Professor, School of Fisheries and Ocean Sciences, University of Alaska
Mr. Eric Gilman, Marine Ecology and Fisheries Specialist, Blue Ocean Institute
Dr. John Halsey, State Archeologist, Michigan Historical Center, Michigan Department of History, Arts and Libraries
Dr. Bonnie McCay, Professor, Department of Human Ecology, Rutgers University
Mr. Melvin Moon, Jr., Director, Quileute Natural Resources Department
Mr. Robert Moran, Washington Representative, American Petroleum Institute
Dr. Steven Murray, Dean, College of Natural Sciences and Mathematics and Professor of Biological Science, California State University at Fullerton
Mr. Michael Nussman, President and CEO, American Sportfishing Association
Mr. Lelei Peau, Deputy Director, Department of Commerce of American Samoa
Mr. R. Max Peterson, Former Executive Vice President, International Association of Fish and Wildlife Agencies
Ms. Barbara Stevenson, Sellers Representative, Portland Fish Pier
Dr. Daniel Suman, Associate Professor, University of Miami
Mr. Thomas Thompson, Executive Vice President, International Council of Cruise Lines
Ms. H. Kay Williams, Member, Gulf of Mexico Fishery Management Council
Mr. Jim Woods, Sustainable Resources, Makah Fisheries Management

Ex Officio Members

Department of Commerce

Ms. Mary M. Glackin, Deputy Under Secretary for Oceans and Atmosphere, NOAA
Dr. Paul Doremus, Acting Assistant Administrator, Program Planning and Integration, NOAA

Department of the Interior

Ms. Kameran Onley, Assistant Deputy Secretary, Office of the Deputy Secretary, U.S. Department of the Interior

Department of Defense/Navy

Designee: Mr. Thomas A. Egeland, Director, Environmental Planning and Conservation Policy, Office of the Assistant Secretary of the Navy (Installations and Environment)

Department of Homeland Security

Designee: LT Jeff Pearson, Deputy Chief, Marine Protected Species, Commandant (CG-3RPL-4), U.S. Coast Guard

Interagency Marine Protected Areas Working Group

Department of Commerce/NOAA

Mr. Joseph Uravitch (Chair), Director, National Marine Protected Areas Center
Dr. Mimi D'Iorio, Geographic Information System and Database Manager, National Marine Protected Areas Center
Ms. Rondi Robison, Conservation Planner, National Marine Protected Areas Center
Ms. Kara Schwenke, Communications Coordinator, National Marine Protected Areas Center
Dr. Charles Wahle, Senior Scientist, National Marine Protected Areas Center
Ms. Lauren Wenzel, Federal Agency Coordinator, National Marine Protected Areas Center
Ms. Heather Sagar, NOAA National Marine Fisheries Service
Ms. Laurie McGilvray, Director, Estuarine Reserves Division (National Estuarine Research Reserves)
Mr. Brad Barr, Senior Policy Advisor, National Marine Sanctuaries Program
Mr. Mitchell Tartt, National Marine Sanctuaries Program

Department of Defense

Capt. Robin Brake, Director, Marine Science, Office of the Assistant Secretary of the Navy (Installations and Environment)

Mr. Thomas A. Egeland, Director, Environmental Planning and Conservation Policy, Office of the Assistant Secretary of the Navy (Installations and Environment)

Ms. Elizabeth Phelps, Marine Scientist, Chief of Naval Operations, Operational Environmental Readiness and Planning

Ms. Lynn R. Martin, U.S. Army Corps of Engineers, Institute for Water Resources

Mr. Joseph Wilson, U.S. Army Corps of Engineers, South Atlantic Division

Environmental Protection Agency

Dr. Brian Melzian, U.S. Environmental Protection Agency, National Health and Environmental Effects Research laboratory, Atlantic Ecology Division

Department of Homeland Security

LCDR Chris Barrows, U.S. Coast Guard, Chief, Fisheries and Marine Protected Species Law Enforcement, U.S. Coast Guard Headquarters (CG-3RPL-4)

LCDR Chris German, U.S. Coast Guard, US Coast Guard Liaison, NOAA, Office for Law Enforcement

Department of the Interior

Mr. Randal Bowman, Office of the Assistant Secretary, Parks and Fish and Wildlife, U.S. Department of the Interior

Ms. Elizabeth Burkhard, Marine Biologist, Minerals Management Service

Mr. Cliff McCreedy, Marine Management Specialist, Natural Resource Stewardship and Science, National Park Service

Mr. Andrew G. Gude, Program Specialist, Refuge Marine Programs, U.S. Fish and Wildlife Service

National Science Foundation

Ms. Roxanne Nikolaus, Ocean Sciences Division

Department of State

Ms. Margaret F. Hayes, Director of the Office of Oceans Affairs, Bureau of Oceans and International Environmental and Scientific Affairs

Dr. Winnie Lau, AAAS Science and Technology Fellow/Marine Science Officer, Office of Oceans Affairs, Bureau of Oceans and International Environmental and Scientific Affairs

Dr. Justin Grubich, Marine Science Officer, Office of Ocean Affairs, Bureau of Oceans and International Environmental and Scientific Affairs

U.S. Agency for International Development

Dr. Barbara Best, Coastal Resources and Policy Advisor, Office of Natural Resources Management, Bureau for Economic Growth, Agriculture and Trade

END NOTES

[1] The purpose of this document is to provide a framework for developing and implementing a National System of MPAs; it is not a blueprint for the establishment of individual MPAs.

[2] Important terms are in bold the first time they are used and defined in the Glossary found in Section VI of this document.

[3] The Department of Defense, the Department of State, the United States Agency for International Development, the Department of Transportation, the Environmental Protection Agency, the Department of Homeland Security, the National Science Foundation, and other pertinent federal agencies.

[4] Adapted from MPA FAC, October 2007.

[5] Given the unique nature of community agreements, whether verbal or written, the requirement for these management agreements to include conservation goals and monitoring and evaluation components may be met through traditional or science-based approaches. In some Pacific Island cultures, for example, management agreements may be part of local oral tradition, and are not written, but would still be considered as meeting this criterion.

[6] A more detailed categorization scheme useful for more in-depth analysis is provided at http://www.mpa.gov.

[7] In most cases, management authority for an MPA lies with one agency or program; however, in certain instances, such as the federal/state National Estuarine Research Reserve System and state/tribe co-management arrangements, authority is formally shared or split among two or more entities. Similarly, Regional Fishery Management Councils have a unique role with the National Marine Fisheries Service in the process for establishing federal fishery management zones and federal fisheries habitat conservation zones. Where explicit agreements and/or legislation govern shared management authority or other formal relationships, the multiple managing entities will be consulted throughout the nomination process. Regional Fishery Management Councils will be a key partner with NOAA in nominating sites to the national system. Through a transparent process, NOAA will consult with its Council partners and fully consider the views and interests of the Councils prior to nominating a site to the national system. These NOAA-Council consultations would take place at the regional-level at key stages of the nominating process, and DOC/NOAA would make final decisions on nominations.

[8] Adapted from: Nigel Dudley and Jeffrey Parish (2006). Closing the Gap. Creating Ecologically Representative Protected Area Systems: A Guide to Conducting the Gap Assessments of Protected Areas Systems for the Convention on Biological Diversity. Secretariat of the Convention on Biological Diversity. Montreal, Technical Series no. 24, vi + 108 pages.

[9] MPA FAC, 2005.

[10] U.S. Commission on Ocean Policy (USCOP). 2004. An Ocean Blueprint for the 21st Century, Washington, D.C.

Chapter Sources

The following chapters have been previously published:

Chapter 1 is an edited, reformatted and augmented version of a Congressional Research Service publication, Report RL32154, updated November 14, 2008.

Chapter 2 is an edited, reformatted and augmented version of a Congressional Research Service publication, Report RL32486, updated October 22, 2008.

Chapter 3 is an edited, reformatted and augmented version of website information from http://mpa.gov/pdf/national-system/finalframework.

INDEX

D

E

F

G

L

M

N

R

S

Y

Z